Developments in Primatology: Progress and Prospects

For further volumes:
http://www.springer.com/series/5852

Paul Naour

E.O. Wilson and B.F. Skinner

A Dialogue Between Sociobiology and Radical Behaviorism

Paul Naour
Central College
Pella, IA
USA
NaourP@central.edu

Series Editor:
Russell Tuttle
Department of Anthropology
The University of Chicago
IL, USA

ISBN 978-0-387-89461-4 e-ISBN 978-0-387-89462-1
DOI 10.1007/978-0-387-89462-1

Library of Congress Control Number: 2008942768

Cover illustration: The photo of E.O. Wilson is courtesy of the photographer, Joe D. Pratt.

Printed on acid-free paper

springer.com

FOR ANN

Man still bears in his bodily frame the indelible stamp of his lowly origin.
(Charles Darwin, The Descent of Man)

The role of variation and selection in the behavior of the individual is often simply ignored. Sociobiology, for instance, leaps from socio- to bio-, passing over the linking individual.
(B.F. Skinner, "Can Psychology Be a Science of Mind"*)*

It is possible that gene-culture coevolution will lie dormant as a subject for many more years, awaiting the slow accretion of knowledge persuasive enough to attract scholars. I remain in any case convinced that its true nature is the central problem of the social sciences, and moreover one of the great unexplored domains of science generally; and I do not doubt for an instant that its time will come.
(E.O. Wilson, The Naturalist)

I regard human evolution as a rapid and ongoing process, made possible by mechanisms loosely described as cultural, which means that human nature will never be set in stone, for better or for worse.
(David Sloan Wilson, Darwin's Cathedral)

But with regard to the material world, we can at least go so far as this - we can perceive that events are brought about not by insulated interpositions of Divine power, exerted in each individual case, but by the establishment of general laws.
(William Whewell, Bridgewater Treatise, *quoted by Darwin on the 1st Edition title page of* The Origin of Species, *1859.)*

Prologue

The history of human behavior, if we take it to begin with the origin of life on earth, is possibly exceeded in scope only by the history of the universe.

(B.F. Skinner, 1981)

"This is going to be a conversation that I will have with B.F. Skinner. This is Ed Wilson. He invited me by to talk about sociobiology. Our relations have always been very friendly and I look forward to it. This should be an interesting talk. So here it is this Thursday morning." Thus begins an intriguing and thought-provoking conversation in Wilson's office between two of Harvard University's most distinguished and accomplished faculty colleagues on a cool November morning in 1987; a date less than three years before Skinner's death. That morning was witness to an intriguing conversation between two of America's most significant intellects – Skinner, recognized for his radical behaviorism that achieved the crowning accomplishment of the so-called behavioral century and E.O. Wilson, acknowledged for establishing, naming and largely developing sociobiology.

The two scientists are bound by their common interest in behavior generally, but more specifically by their inspiration to understand the basis of "human nature" in order to guide humankind to achieve a more productive and sustainable world. They are also bound by the slings and arrows of exaggerated and ideologically based claims against them regarding the "controversial" nature of their work. However, the passage of nearly 20 years since the conversation has certainly diminished, if not altogether eliminated, the harshness of attacks. Yet, challenge (certainly) and outright antagonism (occasionally) appear on a regular enough basis to warrant well-informed evaluation of their work. To be sure, time has placed their work in a far different and more integrated context; a context that is the unifying theme of this book.

I became the fortunate beneficiary of the recording after getting to know E.O. Wilson (Ed) following his 1995 campus visit to a private college in rural Ohio. It was a time during which he was finalizing his conceptualization of consilience and eager to explore the concept with other like-minded scholars. During our early conversations, Ed learned of my early graduate work based in radical behaviorism and later graduate work based in evolutionary psychology. The serendipitous result of our early relationship was his understanding that my background and interest might enable me to identify a unifying thread in the Skinner-Wilson conversation. Thus, I was given a copy of the recording with Ed's invocation to "do something with this someday." That challenge and our occasional correspondence in the years that followed have been an ongoing inspiration to bring the book to its conclusion. The time is right for readers to enjoy the conversation and to be challenged to think how the consilience of radical behaviorism and sociobiology can invigorate a renewal of interest in Skinner and a deeper appreciation for the scholarly achievements of Wilson. In addition, the book is intended to push the boundaries of discipline-based thinking and challenge readers to seek relationships among a wide range of scholarly endeavors in pursuit of a consilient ideal. Those boundaries are challenged in a final chapter devoted to (1) contemporary research that has bearing on present day behavioral science and (2) a discussion of Duane Rumbaugh's rational behaviorism. Finally, an Epilogue by H. Carl Haywood provides both a larger historical context and indications for an important next step toward rational behaviorism.

Considerable effort was given to seeking guidance from others during the production of the book. Their willing help went a long way to guide appropriate corrections and clarifications. However, an inevitable outcome of this considerable interdisciplinary effort is that scholars who mine the ore of singular disciplines will take some exception to details or even portions of the book – perhaps suggesting that some points are wrong or interpretations faulty. If that is the case, perhaps a margin of success has been achieved in guiding some to view the mine from a different perspective in a different light. Wilson challenges us to understand that a "balanced perspective cannot be acquired by studying disciplines in pieces but through pursuit of the consilience among them. Such unification will come hard" (Wilson, 1998). Chapter 3 will especially require his invocation.

This book was produced with the intention that variously sophisticated readers might be drawn for different reasons to its contents. Undoubtedly, some will be drawn to it by virtue of their recognition of one or both of the featured scientists, B.F. Skinner (Burrhus Frederic Skinner) and E.O. Wilson (Edward Osborne Wilson). Hundreds of scholars have spent many years explicating their important work; countless thousands of students have puzzled over the meaning of their theories. Other readers may be drawn to the text because of passing familiarity with the scientific contributions of one or the other – radical behaviorism and sociobiology. Still others will be interested because they are aware of the controversies aroused in the intellectual community by both scholars. I also intend for the book to inspire curiosity among readers who are following the recent work in behavioral neuroscience and asking questions regarding the biological basis for the wide range of behavior considered a part of human culture. Might there be a gene/culture coevolutionary basis to our predisposition to the range of behavior we observe in human courtship practices, marriage customs, and child rearing practices? How might family structures serve a biological imperative? Why is spiritual behavior and its organization into religions so typical among humankind?

Finally, it is time to offer you a more integrated perspective that places the profoundly important work of Skinner and Wilson into a contemporary behavioral context that aligns with the enormous work being done in behavioral neuroscience. Thus, Duane Rumbaugh's concept of emergents along with his rational behavioral perspective are discussed in closing as a means to animate readers from wide-ranging behavioral disciplines to consider how their work might support this integrated view.

Beyond an appeal to those of you having a basis of familiarity and curiosity, my larger objective is to make the book's contents accessible to a wider and, perhaps, less familiar readership in order to provoke renewed consideration of Skinner and Wilson. Undergraduate students of psychology and biology should begin to more commonly encounter topics regarding behavioral psychology and biology in a more integrated way, but they must also be introduced to that synthesis in an uncomplicated and approachable fashion. Thus, the book has been kept intentionally modest in its scope, essentially descriptive in its nature, and purposely related to the larger historical context of Skinner and Wilson, and, more modestly, to the more contemporary context of Rumbaugh. The theoretical aspects of the book are repetitive in places and limited to the primary conceptual contributions of the two scholars – operant conditioning and sociobiology. The deliberate repetitions emerge from different contexts and intentionally guide you to understand the scientific concepts from the differing perspectives that are represented in the book.

Regardless of your motivation, I offer several suggestions before you begin the book and evaluate your competency with the major intellectual contributions of Skinner and Wilson. Those of you who determine that a refresher or quick course might be useful regarding behaviorism and/or sociobiology are encouraged to read Chapters 1 and 2 (Chapter 1 – Selection by Consequences: The Essential B.F. Skinner and Chapter 2 – Human Sociobiology: The Essential E.O. Wilson). Each chapter was written to (1) offer a very quick overview of the major paradigms developed by each scholar, (2) introduce the vocabulary of each paradigm, (3) provide important historical context, (4) offer examples to clarify the paradigms, (5) identify the perceived controversial nature of each paradigm, and (6) set the stage to better engage Chapter 3 – A Consilient View of B.F. Skinner and E.O. Wilson: The Operant Foundation of Sociobiology and Chapter 4 – The Skinner-Wilson Conversation. Another intentional feature of these chapters is a liberal usage of quotations from the various writings produced by both scholars in order to achieve the advantage of an "in their own words" perspective for readers.

You are forewarned that imprecise word choices are maintained when they track the word usage of both Skinner and Wilson. Most notable among those word choices are, "organisms" to represent the more precise: *animals that produce behavior*, and Skinner's customary usage of "respondent conditioning" for what some readers will recognize as: *classical conditioning*. You are also encouraged to make note of Skinner's quick usage of "behaviorist" as he follows Wilson's opening to the conversation. Skinner apparently sought to establish at the outset of the conversation that he and Wilson were similarly disposed when it came to empirical practice. Thus, Wilson is very quick and willing to make it clear beyond any doubt that he accepted Skinner's behavioral approach to all behavior (inclusive of humans). "I'm a behaviorist ... I hold a rigorous cause-and-effect objectivist view of behavior."

This shared perspective was essential to their conversation and, more importantly, to the perception of many readers that two intellects with so seemingly disparate approaches to human nature were able to begin a conversation on that common ground.

Beyond the few idiosyncratic word choices of Skinner and Wilson, you should also take note of the author's regular usage of the proximate cause/ultimate cause distinction along with its more contemporary usage throughout much of the text. The proximate/ultimate biological dualism was advanced by Ernst Mayr (1961) as a means to distinguish between the immediate functional mechanisms (proximate) responsible for driving behavior and the longer-term evolutionary mechanisms (ultimate) driving behavior. According to Mayr (1961), the ultimate mechanisms have been

> *... incorporated into the system through many thousands of generations of natural selection. It is evident that the functional biologist would be concerned with the analysis of the proximate causes, while the evolutionary biologist would be concerned with the analysis of the ultimate causes ... proximate causes govern the responses of the individual (and his organs) to immediate factors of the environment while ultimate causes are responsible for the evolution of the particular DNA code of information with which every individual of every species is endowed.*

Mayr's article provides examples to illustrate the concept of causation but does not push the distinction as a means to understand "kinds of explanations ... It illustrates (but does not name) the principle of explanatory relativity ... The contrast between proximate and ultimate causation is a careful and specific recognition of the relativity of explanation. What looks like the same fact can be given either a proximate or an ultimate explanation" (Amundson, 2005). Sadly, the proximate/ultimate distinction has been most typically expressed as binary and has run the risk of being needlessly biased, rather than viewing the contrast more as a matter of degree. It serves better to view the distinction more logically as a continuum of causality with biological and behavioral mechanisms positioned along that continuum. This concept is expressed throughout the text by intentionally modifying both proximate and ultimate with "relatively" following the custom described so effectively by Sober and Wilson (1998).

Those of you whose confidence with Skinner and Wilson stirs you to explore the "new" content of this book are first encouraged to complete a cursory review of Chapters 1 and 2. The chapters provide historical context leading to the development of their theoretical positions, while also detailing the elements of their theories. Additionally, Chapter 2 includes an important discussion of multilevel selection theory and group selection. No serious consideration of contemporary sociobiology is complete without introducing those critically important concepts as essential contextual features. Although many of you will find the actual conversation inviting at that point, a serious reading of Chapter 3 provides the conceptual framework for the book, generally, and for the conversation, particularly. It is offered as an explication of the conceptual relationship between Skinner's operant conditioning paradigm and Wilson's sociobiology. The chapter also suggests a prospective consilience of their work as revealed in their various writings and their 1987 conversation. Finally, the chapter is intended to push you to consider that conceptual relationship in the larger framework of ultimate and proximate cause in the emergence of social behavior and gene-culture coevolution in humankind – the real basis for the consilience of their work.

Chapter 4 is the edited and sparingly annotated conversation between Skinner and Wilson recorded in November of 1987 at Wilson's request. The two scholars shared decades on the faculty at Harvard University and maintained a warm and positive collegial relationship through those many years. Wilson's invitation to Skinner for a conversation seemed intended to facilitate a "thinking out loud" session to engage him in considering the biological implications of the operant conditioning paradigm. It certainly offers that same challenge to the fortunate reader who comes to the conversation with little understanding of Skinner's biological frame of reference and seeming intent to challenge Wilson to view his life work in that larger context. The edited conversation is intentionally offered with rare interruptions for only the most necessary notes so that readers can fully appreciate the natural conversational flow and inspire a sense for the magnitude of intellectual history brought to bear by the two scholars. The conversation also provides a small window through which to view an historic conversation between two of the most significant intellects of the time. The only experiences missed are the cordial and animated tone of the conversation, the regular clinking of teacups, and the inevitable mishap that caused a slight delay for the necessary cleanup of spilled tea.

Lengthier notes close Chapter 4 and are intended to accomplish three objectives for the reader: (1) bring sequence, closure, and context to our normal conversational style that typically lacks all three, (2) clarify and amplify the many details that are assumed during the conversation, and (3) provide appropriate background information, commentary, and reference as required to guide the reader to better understand the more expansive nature of the conversation than apparent from a surface reading.

The sincerely motivated reader seeking additional insight into this conversation is encouraged to read two of Skinner's most important, thought-provoking but, perhaps, less frequently read publications: "Selection by Consequences," a 1981 article published in *Science*, is essential to the fundamental basis for this book, and the equally important final scholarly work of Skinner's that was literally completed the day before he died – "Can Psychology Be a Science of Mind?" (1990). The latter article is his formal version of the presentation he made just days earlier at the American Psychological Association's annual convention on the occasion of his receipt of its lifetime award. Readers of both Chapter 4 and Skinner's articles will quickly note the alignment of several of the conversational working ideas. It will also become apparent to the astute reader that Skinner was intentionally sharing these ideas with Wilson during their conversation in order to elicit his willingness to think out loud. The more carefully crafted formal version of that thinking emerged as Skinner's final published contribution.

Additional recommended reading includes E.O. Wilson's *Consilience* (1998), an eminently readable book that provides, at least for now, a final synthesis of his comprehensive conceptual framework. The book also provides a well-crafted description of Wilson's theory of gene-culture coevolution. Finally, every reader interested in the essential nature of this book's conceptual offerings is encouraged to read (or reread) Darwin's *Origin of Species* (1859), for its revolutionary insights, its technical merit, and its literary achievement. No better tribute to Darwin's compelling scientific importance exists than Theodosius Dobzhansky's: "Nothing in biology makes sense, except in the light of evolution" (1973).

Chapter 5 is offered to provide readers an overview of writing by Skinner and Wilson that suggests their larger interest – a "concern for the future of the world and its inhabitants" (Skinner, 1989). The theme is woven through the fabric of much of their writing. They had a mutual interest in appealing to readers to deliberately consider the means by which their scientific contributions might make that world a reality.

Finally, Chapter 6 is offered to provide challenge to the reader on two fronts. (1) It overviews current work that might bear on the alignment of Skinner and Wilson. Since their 1987 conversation, there have been significant findings in the neurophysiological research on imitation that promise to advance our understanding of the biological foundation to the operant basis for sociobiology. Fascinating work presently being done on mirror and echo neurons and evolutionary developmental biology (evo devo) are the chief topics of this section of the chapter. The chapter is intentionally written to challenge readers to seek connections among a diversity of research directions with a consilient view of radical behaviorism and sociobiology. Remote as some of the connections may seem to readers, those relationships might invigorate new thinking about the foundations of human nature. (2) The chapter also presents a full account of Duane Rumbaugh's concept of emergents and his rational behavioral perspective. The final section of the chapter is presented to inspire readers to think about an integrated model that provides a structure for including *all* research informing our study of behavior. Rumbaugh's rational behaviorism can connect the (operant) behavioral basis of sociobiology with a perspective that includes cognition and the biological basis for behavior, while it guides behaviorism into the twenty-first century.

Carl Haywood, Duane Rumbaugh's first doctoral student, produced the Epilogue. Haywood skillfully navigates the conceptual nature of the book and then provides a vision for rational behaviorism as "the next big step." I am particularly appreciative that he offers a cogent description of the place occupied by rational behaviorism relative to radical behaviorism and sociobiology within the larger evolutionary framework, suggesting a "kind of sequencing that can provide the springboard to a generation or more of conceptualizing and empirical research." It remains to readers whether you are inspired by Haywood to draw upon that springboard – progressing "from simple to complex explanatory principles, and ultimately from attempts to explain simple and single acts to the explanation of complex, creative, and original thought."

Contents

Introduction

Edward O. Wilson and Julie Vargas

How the Conversation Occurred

I had met Fred Skinner socially on several earlier occasions, with no serious conversation resulting. The contacts were friendly. When he invited me to come by his office in Harvard's William James Hall, I saw an exceptional opportunity to juxtapose our two very different world views in a congenial manner. Even if that were to be accomplished only superficially, it was bound to force out of the two of us some accommodation or, perhaps and just as usefully, a sharp distinction or two, cautiously framed.

Subsequently, he dropped by my office to continue our conversation. With his permission I turned on a small tape recorder, with the understanding that the tape would be kept in an archive and someday be used in historical research. This next step was temporarily delayed when my assistant Kathleen Horton, in a rare mishap for her, mislaid the tape. After several years we presumed that it had been lost, and so largely forgot it. Then, Kathy found it during a spate of office cleaning in a box under her desk, conforming to the rule that diamond rings, hearing aids, and other such valuable objects dropped to the floor seek the most hidden places.

When the tape reappeared, we quickly made multiple copies, one of which Paul Naour transcribed, then added his excellent historical and philosophical introduction. I like to believe that Fred, had he lived to see this publication, would have approved the way Professor Naour has presented both its substance and context.

In retrospect it is clear that when we had this conversation, the understanding of human behavior was in rapid flux. At that time a great many – perhaps most – social scientists and philosophers still viewed the human brain as essentially a blank slate, in accordance with radical behaviorism. There was, in this view, no human nature. On the other side, the attempt to reinterpret human behavior through animal ethology and sociobiology was still a bit simplistic, and evidence pointing to a genetically based human nature was sparse and fragmented. Skinner was open to some degree of genetic programming in learning predisposition, and I certainly could agree that learning was a major element in the unfolding of most genetic programs.

The two approaches were on their way to a badly needed degree of complementarity. The past 22 years have seen just such a synthesis emerge, and with increasing clarity, through studies in neuroscience, human genetics, developmental cognitive psychology, and sociobiology (the latter usually called evolutionary psychology when applied to humans). Still, the human mind and human nature remain substantially an open frontier for researchers in both the natural and social sciences.

A Daughter's Commentary

My father, B. F. Skinner, had a very high regard for E. O. Wilson both as a scientist and as a person. The two men had much in common. Skinner's discoveries, like those of Wilson, had arisen from direct observation of life processes in action. Both scientists got their hands dirty. Neither was an armchair philosopher, but both saw the implications of the scientific principles they were observing. Both men wrote articles and books to warn about the consequences of the careless actions mankind is taking. Going beyond warnings, each proposed solutions for creating a more humane and sustainable world, and each lived the principles he espoused. Their solutions for rescuing the world lay in different fields, but their concerns overlapped.

While Skinner was still alive, Wilson worked in an historic building that housed Harvard University's famed glass flower collection. He had a large office with a high ceiling characteristic of older buildings. Wilson's office was lined with books, but its main feature was a three yard long colony of leaf-cutter ants. A visitor was offered a stick and a suggestion to gently poke the ants' edifice to watch the soldier ants come out to defend their colony. With little prompting, Wilson would take a visitor to rooms filled with tall dark brown cabinets of narrow wooden drawers that housed Harvard's archival collection of ant specimens – the largest collection in the world.

Skinner's office was on the seventh floor of William James Hall, a high rise modern building. The office was not large. It held books along one wall and, for many years, a philodendron plant arising from a ten-inch pot on the shelf under the window. As the plant grew, Skinner fastened string up the wall and across the ceiling until at one point the main strand cross-crossed the ceiling several times. Standing under the plant's tendrils, a visitor could hear the clacking of apparatus from the lab down the hall. A visitor showing any interest in research would be offered a tour of the lab. The window in Skinner's office overlooked the building less than a block away that housed the glass flowers and Wilson's office.

Though the two men were so close physically and philosophically, they rarely spent time together. Each scientist had created a field that demanded continual attention: sociobiology for Wilson and behaviorology for Skinner. He used the name "the experimental analysis of behavior." Since the origins of the two fields in the twentieth century, the fields have progressed in their own spheres with as little interaction between them as between their founders. Increasingly that isolation is being called into question. It is now clear that all life forms on earth are interconnected and that behavior, human or ant, is a product of interactions of life forms within their world. Even the genes that are "turned on" during development depend on environmental variables that, in turn, are affected by the behavior of each species. Further consilience between the biological and behavioral sciences would advance both.

This book documents the only recorded discussion known between E. O. Wilson and B. F. Skinner. It shows agreement between the two scientists that intention has no place in a science of behavior or in evolution. Skinner's definition of positive reinforcement, for example, is an event that increases the rate of a class of actions on which it is made contingent. Conceiving of a positive reinforcer as a "pleasant consequence" misleads the teacher whose scolding of a student's swearing *increases* his swearing. Other misconceptions of Skinner's basic position are common. It's not clear who first talked of a "black box" between stimulus and response. It was not Skinner. Skinner's point about looking inside the brain was that neurophysiological events occupied a different domain from the analysis of behavior. The neuronal activities going on inside my head as I type the words you are reading are useless for explaining why I wrote what I did. For that you need an analysis of contingencies, the same contingencies, in fact, that are responsible for whatever brain activity occurs. Neurophysiological events do not explain the superorganism status of ants, either.

Both Skinner and Wilson describe relationships based upon what they observed and inferences tightly based on those observations. They advanced more than a science. Both men propose a world view ultimately concerned with the fate of the human species and the world in which all life forms live. The conversation featured in this book gives a unique look into their scientific and humanitarian goals.

Chapter 1
"Selection by Consequences": The Essential B.F. Skinner

In a scientific analysis, histories of variation and selection play the role of the initiator. There is no place in a scientific analysis of behavior for a mind or self.

(B.F. Skinner, 1990)

This book is about human nature – more precisely, human behavior and culture from both their overt foundation in "real-time" environmental contingencies (Skinner) and their material basis embedded in the long-term (socio) biological processes of human evolution (Wilson). This scientific perspective is considered by many to deny the phenomenological concept of mind. Thus, by necessity the point of view that is most apparent and prevails throughout much of this book is that a self-determining mind independent of environmental contingencies is not a viable feature of radical behavioral and most biological approaches. However, you should also make special note of the fact that the final chapter is offered to show you the way to a more "rational behavioral" (Rumbaugh et al., 1996), less dogmatic, and less polarizing perspective. The prospect of that more "emergent" concept of mind augurs well to reconcile the long-lived bipolar study of human nature represented by Skinner and Wilson and move the study of human nature to a far more integrated position.

The conversation between B.F. Skinner and E.O. Wilson animates the bulk of this book, while also inspiring its closing view to the future. Thus, human behavior and culture are mostly addressed from the empirical point of view of these two leading forces in twentieth century American intellectual life. In particular, their wide-ranging work had significant influence on framing the scientific conversation regarding human nature over the past 50 years, while continuing to stimulate important scholarly work and intellectual controversy (Pinker, 2002). A better understanding of their work may well offer a clearer picture of where we go from here.

You will benefit more from this book by reviewing the scientific paradigms and language of both Skinner and Wilson prior to proceeding to the conversation and the remainder of the book. This review will also more importantly provide the foundation for a better understanding of the fundamental relationship of their work and the prospect for their consilience. Thus, Chapters 1 and 2 provide the requisite background for you to establish the foundation to better understand the focus of

P. Naour, *E.O. Wilson and B.F. Skinner*, Developments in Primatology: Progress and Prospects, DOI 10.1007/978-0-387-89462-1_1,

Chapter 3 on a consilience of their work and to set a larger context for their conversation that is the basis for Chapter 4.

The Behavioral Century

Twentieth century American psychology is often characterized as the century of behaviorism. Without question, the most prominent figure in that century was B.F. Skinner. The major portion of this chapter will provide an overview of Skinner's operant conditioning paradigm and the technical terminology he created to identify its many components. However, the immense influence he had on shaping that era did not come in isolation or without significant contributions of many important predecessors. Thus, several key individuals, their work, and the relationship of their work to Skinner's new paradigm will precede the chapter's major emphasis. The initial portion of the chapter is not intended as a comprehensive overview of the history of behaviorism. Rather, it will focus only on three key persons responsible for the foundational work necessary to understand the context of Skinner's operant conditioning – Ivan Pavlov, John Watson, and Edward Thorndike.

The stage was set for Skinner's mid-twentieth century emergence by Ivan Pavlov in the late nineteenth century with his empirical description of respondent conditioning (Skinner's custom to utilize respondent conditioning as his phrase of choice will be respected in this text; readers should be reminded that it is fully interchangeable with Pavlovian conditioning or, more commonly, classical conditioning). Pavlov was intrigued by the serendipitous observations he made of salivation behavior in dogs as he studied their digestive systems. He observed that dogs salivated at the mere sight of caretakers as they arrived to prepare for feeding the animals. Presumably, the animals associated the caretaker with the prospect of food, which elicited their involuntary salivation. Pavlov set out to empirically determine the basis for this observation. It is this empirical work, developing the respondent conditioning paradigm, that earned his place in the history of science. Conversely, it was his primary work on the physiology of digestion that earned him the 1904 Nobel Prize in Physiology and Medicine.

The importance of Pavlov's paradigm was magnified by the coincidental disintegration of confidence in mentalistic approaches to understanding brain and behavior relationships, best represented by phrenology. The rejection of the work attempting to catalog relationships between indirect measurements of brain and behavior resulted from the increasing awareness that the data were spurious and often reinforced the worst social stereotypes of the day. The emerging discipline of psychology was ripe for a paradigm that was firmly based in best empirical practice. Pavlov's respondent conditioning paradigm provided scholars both observable and quantifiable behavioral variables that enabled them to recapture this solid empirical footing. Thus, it became the experimental basis for many academic psychologists advancing the emerging field of psychology – most particularly in American psychology.

Most notable in Pavlov's paradigm is his description of the means by which a neutral stimulus can be paired with a causal stimulus to produce an involuntary response. The procedure was described by Pavlov as producing a conditioned reflex – more commonly known today as "conditioning." A neutral stimulus is defined as one that has no history of producing a response in the organism when presented in isolation. However, once there are a number of closely contiguous pairings of a neutral stimulus with a causal stimulus, the organism will be "conditioned" to produce the involuntary response in the presence of the neutral stimulus alone. The classic example is Pavlov's experiment in which he demonstrated that he could "condition" a dog to salivate to a previously neutral stimulus – ringing a bell. Pavlov paired the bell closely in time (seconds) with presentation of food. Several repetitions of pairing the bell and food resulted in a circumstance where the causal food stimulus was withheld and the dog salivated to the bell presented in isolation.

Pavlov's original serendipitous observation was verified. Several pairings of the previously neutral appearance of a caretaker linked with the closely contiguous presentation of food quickly "conditioned" salivation in the dog. The mere appearance of the caretaker elicited the salivation behavior. In both examples, neither the bell nor the caretaker would elicit salivation until paired with a causal stimulus. Simple as these observations may seem, they provided a major advancement in behavioral science and were responsive to the primary requisite for empirical practitioners to precisely describe observable and quantifiable variables as they cultivated their new science of psychology. It is of some relevance to note at this point that although it is popularly believed that a bell was used as Pavlov's neutral stimulus, historical evidence actually suggests that he used a variety of stimuli, including whistles, metronomes, and tuning forks, along with various visual stimuli.

Experimental manipulation quickly proceeded on the variables described in respondent conditioning and technical terminology emerged to provide a common language among investigators. That terminology is important in the larger framework of behavioral psychology, in that misapplication of terms led to misrepresentation of Pavlov's original paradigm and misunderstanding of Skinner's operant conditioning paradigm as it emerged mid-century. Thus, the following terminology defines classical conditioning. The stimulus that elicits an involuntary response is called an *unconditioned stimulus* (UCS); its presentation unequivocally causes an organism to produce an *unconditioned response* (UCR). For example, food (UCS) will always cause an organism to salivate (UCR), just as fear will always cause an organism to have accelerated heart rate – among other involuntary physiological responses. In both cases, should the UCR be contiguously preceded by an unrelated *neutral stimulus* (NS) on several sequential occasions, that neutral stimulus will be conditioned to elicit what is now identified as the *conditioned response* (CR) when presented, absent the UCS. Therefore, once the neutral stimulus elicits the involuntary response, it becomes a conditioned response (CR) in the respondent conditioning paradigm. It bears noting that the CR is never identical to the UCR, in that the formerly neutral stimulus that emerges as the conditioned stimulus (CS) elicits only a part of the original response elicited by the unconditioned stimulus. An important consequent of the "practice" required to maintain the UCS/CS – CR relationship is the predictable variance in the environment from one "practice" to the next. Thus, the topology of each CR inevitably varies (Guthrie, 1952).

By way of two examples, the classical conditioning paradigm can be illustrated as follows:

Example 1

	food (UCS)		***causes***	salivation (UCR)
bell (NS)	+	food (UCS)	***causes***	salivation (UCR)
bell (NS)	+	food (UCS)	***causes***	salivation (UCR)
bell (NS)	+	food (UCS)	***causes***	salivation (UCR)
bell (CS)			***causes***	salivation (CR)

Example 2

		harsh teacher (UCS)	***causes***	fear response (UCR)
school building (NS)	+	harsh teacher (UCS)	***causes***	fear response (UCR)
school building (NS)	+	harsh teacher (UCS)	***causes***	fear response (UCR)
school building (NS)	+	harsh teacher (UCS)	***causes***	fear response (UCR)
school building (CS)			***causes***	fear response (CR)

Example two is a common description of the means by which school phobia can be classically conditioned in young children. In a circumstance that was previously neutral (the sight of an elementary school building from the window of a school bus), respondent conditioning can elicit a fear response in a child who experiences a harsh tone from a teacher contiguous with their experience of sighting the school from their school bus. This particular experience in children on just a single occasion is powerful enough to respondently condition that fear response at the sight of a school building for many years.

All cases of respondent conditioning are contingent on the response being completely *involuntary* or as Skinner describes them, "responses prepared in advance by natural selection," such as digestion, respiration, and reproduction (Skinner, 1981).The emphasis on involuntary is intended to underscore the most frequently misunderstood aspect of respondent conditioning; it describes conditioning of behavior that are exclusively controlled by smooth muscle or glandular responses – in other words, behaviors that have an absolute causal relationship with a stimulus over which an organism has no control. Most importantly, Pavlov demonstrated that it was possible in respondent conditioning for those responses to come under the control of new (neutral) stimuli and a means by "which individual organisms acquired behavior appropriate to new environments" (Skinner, 1981).

Pavlov's work provided early behaviorists with an array of variables they could manipulate but none that enabled them to investigate the far larger class of voluntary behavior. Nonetheless, the emerging discipline of psychology was presented with an opportunity to assert its positivist approach and more significantly unify psychology with natural science. Respondent conditioning's limitation to a relatively small class of involuntary behavior did not limit the ambitions of investigators to creatively advance Pavlov's paradigm and broaden its scope to include voluntary behavior as well. Most notable among those individuals in the early twentieth century were John Watson and, to a lesser extent, Edward Thorndike.

Watson's influence on bringing a strident behavioral approach to psychology often garners him the title, "father of American psychology." His 1913 publication in the *Psychological Review* of the article "Psychology as the Behaviorist Views It" is frequently referred to as the behaviorist's manifesto. In it, he identifies the necessity to observe only overt behavior, to eliminate introspection as a means to describe behavior, and to offer interpretations of behavior not dependent on subjective states of consciousness – "Psychology is a purely objective experimental branch of natural science. Its theoretical goal is prediction and control of behavior." Thus, psychology for the first time is identified as a science, and behaviorism becomes the philosophy of that science (Skinner, 1989).

Watson aims solely at overt stimuli, responses, and the observable variables that might have an effect on an organism's behavior. Watson championed reductionism, advocating for empirical understanding of environmental stimuli for their effect on observable behavior. His work is also noteworthy in laying the foundation for understanding animal behavior as an evolutionary continuum across species. This empirical focus results in more firmly establishing ethology as a scientific endeavor aimed at causal descriptions of animal behavior, while earning that discipline scholarly credibility.

Beyond his ambition to clearly define a behavioral approach for psychology, Watson framed the major categories of behavior that served as the basis of behaviorist descriptions for years – explicit, implicit, hereditary, and acquired. The precision of his behavioral focus achieved even more prominence with the beginning of World War I and the subsequent demand for objective means to assess human ability and provide training to enhance performance. A uniquely American orientation to psychology was affirmed and established, a bearing that would dominate most of twentieth century American psychology. Mentalism no longer appeared in that picture.

Edward Thorndike, a contemporary of Watson, is equally important to the emergence of behaviorism as the basis of American psychology. His work, though lesser known than Pavlov's to the wider public, is nonetheless more significant as the basis for Skinner's work yet to come. Thorndike's most significant conceptualizations regarding behavior came as a result of careful experimental studies of animals. Thus, he was able to achieve a far greater measure of experimental control in his research. His most commonly known and noteworthy findings were derived from his studies with cats. Thorndike developed novel latching mechanisms for cages that cats were able to manipulate in order to achieve their freedom and have access to food. Inevitably, so long as the behavior was within the repertoire of possibility for the cats, they achieved release. Careful observation of the cats' behavior led Thorndike to suggest that they were able to form an association between a stimulus (S) and a response (R) – presumably their hunger and food availability serving as the stimulus and the manipulation of the locking mechanism as the response (R).

Thorndike's observations led him to propose that the cat was instrumental in producing a response that led to a satisfying consequence. He suggested that the founding principle of "instrumental learning" was the *law of effect*. In other words, an organism produces a variety of behaviors in order to achieve a satisfying result (effect). Pleasurable results for the cat lead to stamping in the stimulus–response relationship, while unpleasant results stamp out the stimulus–response relationship. The phenomenon that Thorndike observed wherein an organism produces a wide variety of behaviors in order to achieve a satisfying result was identified by him as the *law of multiple responses* (more easily defined for some as trial and error behavior).

These two laws are critical advancements in behavioral psychology, regardless of how simple and intuitive they may seem. Unfortunately, Thorndike did not conjecture that instrumental learning and his two laws are innovative advancements that extend the standard S–R paradigm. The law of effect distinguished a focus clearly beyond the momentary stimulus–response relationship for the first time to include consequences that resulted from responses produced by the organism. There was no specific causal relationship apparent in the organism's behavior. In reality, Thorndike's law of multiple responses identified that the organism was "instrumental" in producing various context-appropriate behavior. Such trial and error-like behavior suggested that organisms are not the S–R machines that critics of behaviorism suggested; rather, organisms actively sought to produce behavior that achieves pleasant consequences.

Behaviorism enjoyed immense popularity among academic psychologists through the first half of the twentieth century. Applications were most common in education, counseling and therapy, training for military purposes, and manufacturing. Many creative ways were identified to explore the wide array of behavioral variables in animal and human performance. As well, some investigators described systems that packaged their new behavioral approaches. Among the more commonly known was Edwin Guthrie. Guthrie described one-trial learning and habit formation. He suggested that one S–R pairing was enough to establish an association and produce a habit that will recur in the presence of the stimulus. Repetition of the S–R pairing, otherwise known as practice, only serves to more firmly stamp in the habit. Guthrie is also remembered for identifying several behavioral means by which to break bad habits that might have been stamped in by pairing. These habit-breaking techniques are still in use today by behavioral therapists.

Many other psychologists contributed in significant ways to the behavioral century, each cumulatively advancing the development of the foundation for a new behavioral paradigm. However, none had offered a theoretical departure from the behavioral foundation established in respondent conditioning by Pavlov. Thus, the behaviorist paradigm through the first half of the century was limited to the causal relationship between stimuli and responses, involuntary responses of organisms, and the physiological mechanisms of reflexes. The intellectual insight and courage to alter the existing paradigm was left to B.F. Skinner.

Skinner's New Paradigm

Both natural selection and operant conditioning have been slow to make their way as scientific explanations because they conflict with well-established views.

(B.F. Skinner, 1989)

Skinner seized the opportunity made available by his predecessors and quickly enlarged the behavioral influence on American psychology. His radical approach to behaviorism extended their foundational work, particularly Thorndike's paradigm-challenging observations. He developed a new conceptual framework, broke the standard S–R paradigm, and developed a comprehensive new system that he named *operant conditioning*. *The Behavior of Organisms* (1938), Skinner's first book, established the trajectory of his intellectual life. It came as a result of his graduate school experiments at Harvard and time as a research fellow. This rich intellectual early period for Skinner prompted his development of the operant conditioning paradigm and its many constituent components – each with a wealth of experimental opportunities. Additionally, he developed the famous Skinner Box during this period as a means to address his disregard for the unsophisticated nature of experimental instrumentation available at the time. Its elegantly simple technology precisely quantified behavior and allowed for total control of experimental variables; the Skinner Box is utilized to this day.

Skinner's groundbreaking work was motivated by his frustration with the limitations of behavioral investigations focused on reflexes, "I wanted to study the behavior of the organism as a whole ... Conditioning involved a certain amount of 'prediction and control,' but it was the behavior of an organ, not an organism"(Skinner, 1989). Skinner elegantly recapitulates in that same 1989 publication his experiment that achieved for him the intellectual breakthrough that resulted in operant conditioning. In his own words,

I built an apparatus in which a white rat ran along a delicately mounted pathway. The forces exerted on the pathway were recorded more or less as Sherrington had recorded the forces exerted by a single muscle on his 'torsion wire myograph.' But something else turned up in my experiment. The rat was hungry and got a bit of food at the end of each run. I noticed that after it had finished eating it did not always start immediately on another run. The delays in starting seemed to vary in an orderly way, and that suggested another kind of control in the 'organism as a whole.' After a long series of steps, I found myself recording the rate at which the rat ate pellets of food or got them by pressing a lever.

I recorded the behavior in a cumulative curve, a form of graph not well understood for a long time. It had many advantages. The curve that resulted showed a steady decline in slope, suggesting an orderly process of satiation as the rat ate its daily ration. When I did not let the rat get pellets for a few minutes, it ate more rapidly when they were again available, and the cumulative record rose to meet a rough extrapolation of the earlier part. A rather subtle change in behavior was exposed to view.

... In some sense the lever must be acting as a stimulus, but I could not turn it off or measure it. The rate at which the rat ate pellets of food or pressed the lever could serve as an alternative, however. Rate of behavior has proved, in fact, to be a highly useful dependent variable.

... It was such a unit (of discrete behavior – the press of the lever) that I called an operant. What was reinforced was a response as an instance; what was strengthened was an operant – the probability that other responses would occur.

> *... What remained to be done in a science of behavior seemed clear. I should look for other independent variables and observe their effects.*

Skinner redefined the behavioral focus to the *consequences* of behavior from the former overriding and primary concern regarding the association between stimuli and response. Antecedent stimuli were no longer of primary interest. Rather, the major focus was placed on stimuli that are made available *following* behavior emitted by an organism. Skinner called that basic unit of behavior an *operant*. The term is rooted in the word "operate" and is intended to evoke the image that organisms "operate" on their environments by emitting behaviors that are more likely to achieve favorable outcomes. Emitted operant behavior is only important to Skinner insofar as it relates to empirical measures that provide evidence for effect. Thus, the result he identified as the primary experimental datum of operant conditioning is the organism's *rate of behavior*. "According to my experiments, what happened after an organism behaved played a much larger role than what happened before" (Skinner, 1989).

The Vocabulary of Operant Conditioning

The technical terminology of Skinner's radical behaviorism is critically important to achieve a clear understanding of operant conditioning. Most of the terminology was newly identified as he conceptualized the paradigm and was appropriately descriptive of the various principles. However, Skinner did choose to utilize common behavioral terminology to represent the basic paradigm by continuing the custom of utilizing both the *S* (stimulus) and *R* (response) symbols of earlier behaviorists. Many newcomers to operant conditioning are initially confused by his usage of the *S* and *R*, in that he seems to reverse the order of their flow and he clearly does not want them to represent a causal relationship. The significant departure from the S–R terminology of the respondent conditioning paradigm that preceded Skinner is most notable in his reversed ordering of the R–S, an indication of a clear shift away from the S–R causal paradigm.

Skinner initiated his new paradigm with *R* to represent an organism's free operant responding in the environment. In other words, Skinner's *R* indicates that organisms "freely" emit behavior as they operate on their environment. The organism's free operant responding/behavior (*R*) will become more or less likely to recur as a result of what occurs in the environment as a consequence. The consequence of which Skinner speaks is some stimulus (*S*) made available in the environment that affects the likelihood the organism will repeat the behavior. Those consequent stimuli (*S*) are naturally occurring in the environment or purposely provided by someone intending to change the rate of a target behavior. Consequences that are described as *reinforcing* always increase the likelihood that the behavior (*R*) will recur. Thus, the symbolic representation of the paradigm is most typically seen as: $R - S^{R}+$.

For example, a newly hatched robin thrusts its neck (R) farther than its nest mates and achieves the consequent of the mother's food bits (S) first. This pleasant reinforcing consequence (S^R+) makes it more likely that the infant robin will emit neck-thrusting behavior again. An eagle is likely to fly (R) the narrow free-flowing water of a river in midwinter to feed (S^R+) as the fish swim in unfrozen open water to achieve oxygen, or athletes will give enormous time to preseason training routines (R) with the prospect of achieving a place on the team (S^R+). *R* represents the emitted behavior/response and Skinner's S^R+ represents the pleasant (+) stimulus (S), which made more likely that the behavior would recur by reinforcing (S^R+) it.

A more subtle, yet more powerful, means to *increase* behavior is negative reinforcement. An eagle is likely to fly (R) the narrow free flowing water of a river in midwinter in order to avoid hunger (S^R-) as the fish swim in unfrozen open water to achieve oxygen. Athletes will give enormous time to preseason training routines (R) in order to avoid the prospect of not making the team (S^R-). Any consequence to a behavior that is likely to increase the probability of its recurrence is characterized as a reinforcer. In this case, the increase in behavior is brought about by the prospect that a negative stimulus will be present if the behavior is not emitted. Most behaviorists suggest that negative reinforcement is the most common basis for the increased probability of appropriate behavior. The majority of humans behave (R) in socially appropriate ways in order to avoid the prospect of being rejected (S^R-) by peers or punished by eternal damnation. The punishing consequent is not played out – it is the prospect for the punishing consequent. Some behaviorists prefer suggesting that negative reinforcement is better described as withholding a positive reinforcer. However, it more frequently achieves its most effective result and more common application when cast in threatened unpleasant consequences. Indeed, when utilized in its most crudely manipulative manifestation – torture and brainwashing – the individual is intentionally placed in a negative environment and offered the prospect for escaping that environment by emitting behavior the manipulator desires.

Skinner was quite emphatic about the ethics of using only positive reinforcement to increase behavior. He maintained that negative reinforcement could easily be manipulative and abusive. Further, Skinner urged only the rarest application of punishing consequences in order to reduce or eliminate undesired behavior – when an individual could potentially hurt themselves or others. When necessary, positive punishment (S^P+) is a stimulus presented following a harmful behavior as a consequence designed to diminish the behavior. The common spanking is the best example. Negative punishment (S^P-) is the removal of a stimulus the individual would like to retain as a consequence to their undesired behavior. A fine imposed following a speeding violation and imprisonment following a more serious crime are both examples of stimuli we all wish to maintain – money and freedom.

In summary, the operant conditioning paradigm and its four consequences are as follows:

Positive Reinforcement
Patient vigilance at water hole (R) → prey appears (S^R+)
Quiet attentiveness in class (R) → verbal praise and extra recess (S^R+)
In each case, the behavior will likely increase because a pleasant consequence was added.

Negative Reinforcement
Avoid the water hole in small group (R) → avoid attack by predator (S^R−)
Driving under the speed limit (R) → avoidance of a speeding ticket (S^R−)
In each case, the behavior will likely increase because unpleasant consequence was avoided.

Positive Punishment
Break a neighbor’s window (R) → receive a scolding (S^P+)
Throw food off of high chair tray (R) → receive a spanking (S^P+)
In each case, the behavior will likely decrease because an unpleasant consequence was added.

Negative Punishment
Break the speed limit (R) → pay a fine and court costs (S^R−)
Perform an armed robbery (R) → lose freedom (S^R−)
In each case, the behavior becomes less likely to recur because a valued stimulus was removed.

Reinforcers, Cues, and Operant Shaping

Natural selection prepares an organism only for a future that resembles the selecting past. That is a serious limitation, and to some extent it was corrected by evolution of a process through which a different kind of consequence could select additional behavior during the lifetime of the individual. The process is called operant conditioning and the selecting consequence a reinforcer.

(B.F. Skinner, 1989)

The nature of reinforcement, behavioral acquisition, and behavioral maintenance became Skinner’s primary focus once he identified the four basic forms of stimulus consequence to emitted behavior. Reinforcers are considered to be primary – necessary to sustain life – or secondary – their reinforcing value must be conditioned by initial pairing with a primary reinforcer, followed by gradual withdrawal of the primary reinforcer. Food, oxygen, water, sleep, contact comfort, and attention all provide for a primary reinforcing effect. However, such reinforcers should never be used as a means to induce the acquisition of new behavior or to sustain behavior.

Skinner long-maintained that humankind has a right to freely access primary reinforcers and that effective behavior management always means pairing primary reinforcers with stimuli that can be conditioned to have a reinforcing (secondary) effect. When a parent offers contact comfort and pairs that with a positive verbal utterance, a child is quickly conditioned to those words; over time, the words can substitute for the actual contact comfort. In short order, words become powerful as secondary reinforcers to condition new behavior or to sustain formerly acquired behavior. It is not long before children begin to identify to parents their secondary reinforcement preferences – playtime, TV, video games, and money. In fact, the reinforcing value of money is amplified by its generalized nature. The actual paycheck, cash, or coins have no real merit as reinforcers. We must be conditioned to their generalized secondary reinforcing effect. Once conditioned, they can be exchanged by the recipient for all manner of things over which they have the privilege of choice, including primary reinforcers. Unquestionably, money, as a generalized secondary reinforcer, achieves significant reinforcing effect for much of our lifetime.

Obviously, an important consideration regarding effective reinforcement technique is the strategic presentation of the reinforcer. Schedules of reinforcement and their effect on behavior acquisition and behavior maintenance were significant variables for long-term investigation by radical behaviorists and their many students (Skinner, 1953; Ferster and Skinner, 1957). How can reinforcers be utilized to maximize behavior acquisition? How can reinforcers sustain behavior and then resist extinction over long periods of time? What schedules of reinforcement produce the highest quality behavior and achieve the greatest resistance to extinction? Such questions were the stuff of countless investigations and in thousands of real-world behavioral management applications.

Reinforcers are identified as either continuous or intermittent. Continuous reinforcers are offered each time the behavior occurs and are best utilized to initially condition a behavior. A very early consequent of continuous reinforcement is the probable extinction of the behavior when the reinforcer is not made available for a period of time. Thus, once a behavior is conditioned under a continuous schedule, the reinforcer should be made available less frequently by utilizing an intermittent schedule of reinforcement. Intermittent schedules can be systematically managed by employing an interval schedule or a ratio schedule. Interval schedules progressively increase the interval of time between reinforcement availability, contingent upon the individual performing the behavior. Ratio schedules gradually increase the actual number of behaviors expected before the reinforcer is offered. Obviously, the reinforcer must be offered both contingently and contiguously with the performance of the behavior.

Both interval schedules and ratio schedules can be on a fixed or a variable system of presentation. A fixed interval schedule identifies a specific period of time between reinforcement availability, while a variable interval identifies an average period of time between reinforcement opportunities. Similarly, fixed ratio schedules establish a specific number of behavior events necessary before reinforcement and variable ratios identify an average number of behavior events prior to reinforcement. The key to effective application of the interval schedules is to incrementally increase the interval or ratio in a gradual fashion. Not doing so may introduce unnecessary reinforcement stress and risk extinguishing the behavior.

Significant empirical evidence demonstrates that each of the interval schedules has very specific outcomes related to the rate of response (RR), quality of response (QR), delay in responding following reinforcement, or postreinforcement pause (PRP), and resistance to extinction (RE). A very simplified summary of those outcomes is offered below with examples that follow. Far more detailed descriptions are provided in a variety of Skinner's publications.

An intriguing reinforcement phenomenon that has been well documented in many investigations is described as vicarious reinforcement. A behavior may become more likely to occur from an individual who simply observes others being reinforced for producing that behavior. This obviously occurs among humans, but can also be observed

	RR	QR	PRP	RE
Fixed Interval	mixed	mixed	significant	low
Variable Interval	high	steady	none	high
Fixed Ratio	mixed	mixed	significant	low
Variable Ratio	high	steady	none	high

Fixed interval: the standard paycheck and regularly scheduled classroom tests
Variable interval: surprise quizzes or random attendance checks
Fixed ratio: a set number of homework problems or piecework manufacturing
Variable ratio: slot machines or a base hit after any number of pitches

in primates. Strategic use of vicarious reinforcement is an effective way to model appropriate behavior and demonstrate a favorable outcome. This phenomenon is particularly interesting in light of very recent neurological research that has identified the existence of mirror neurons and echo neurons. These neurons are similarly active whether they are responsible for the actual production of the behavior or simply observing it. The evolution of this biological attribute may go a long way in explaining the natural selection of operant conditioning as a means for organisms to develop new and environmentally responsive behavior – Skinner's "second kind of selection by consequences." Additional discussion of this concept and its relationship to modeling and imitation is offered in Chapter 6.

Another means by which individuals may be guided to produce behavior is to provide a discriminative stimulus that primes (cues) the individual. Verbal cues are the most common examples. Skilled teachers can be observed to utilize more apparent multiple cues to set the stage for appropriate behavior and contingent reinforcement. Teachers will often combine verbal cues with hand gestures to "coach" the behavior they seek. More subtle cues include the wide range of social cues we seek as we negotiate the social world in which we live. Attention and acceptance by others are powerful reinforcers; observing others for cues and behaving accordingly is likely to achieve reinforcement, particularly vicarious reinforcement. Certainly, these discriminative stimuli, cues, appear to be ubiquitous and warranted for Skinner their own symbolic representation – S^D. Thus, the following might represent a second grade classroom in which a teacher is cuing quiet behavior:

A quick off/on of the light switch (S^D)
Students are quieted by the startle (R)
"Thank you" (S^R+)

So effective are cues that they also provide us information regarding behavior that might lead to punishment if we produce them. Skinner identified these cues with the symbolic representation S^Δ. In other words, an S^Δ is a discriminative stimulus that warns the individual against doing something – or pay the unpleasant consequence. The visual cue of a police car parked in the median along a highway is a powerful S^Δ that cues a driver to slow down or risk a speeding ticket.

Simple and discrete behaviors that are the fare of typical textbook examples are far less common than the more complex behaviors we seek from ourselves in the real world. How do we achieve complex multidimensional behavior when it is unrealistic to expect those sophisticated terminal behaviors to be operantly emitted? Complex behavior simply will not occur within the normal behavioral repertoire of organisms. You can't reinforce what doesn't exist. Skinner identified a highly effective operant-shaping technology in response. The effective behaviorist identifies distinct approximations of that final behavior and strategically reinforces those achievable elements of the expected outcome. Operant shaping toward the terminal behavior results from gradually increasing the contingency of reinforcement to successively approximate the terminal behavior. Ultimately, the final behavior can be achieved with patience and vigilance to the technology of operant shaping. Lassie must first be conditioned to attend to her trainer before she is reinforced for approaching a light switch on verbal command. There is considerable time and many intervening approximations that must be shaped before Lassie stands against the wall and flips on the light switch with her extended paw.

Beyond a Technology of Behavior

Behaviorism was just 15 years old when Skinner arrived on the Harvard University campus in 1928. It was a short 10 years later that he turned his research into *The Behavior of Organisms* – the basis for radical behaviorism. A review of his comprehensive empirical record generated over the next five decades reveals an enormous and, perhaps, unrivaled intellectual energy. Beyond the comprehensive empirical detail regarding all aspects of operant conditioning, he was equally committed to encouraging general application of the science of behaviorism to guide our better understanding of human behavior and to ultimately provide the behavioral foundation to create a better world. In order to do so, he often found it necessary to confront and challenge prevailing wisdom regarding mentalistic notions, pseudoscientific principles of behavior, and ineffective social customs. He was unfailing in his willingness to do so and to receive the, oftentimes uninformed, attacks of his critics.

Walden Two (1948), Skinner's utopian novel, was his first serious attempt at offering real-world application of behaviorist principles to create an empirically designed community in which "... governmental, religious, and capitalistic agencies are replaced by face-to-face personal control. New members begin by following simple rules, with the help of instruction and counseling, and their behavior is soon taken over by carefully designed social contingencies" (Skinner, 1989). The utopian society makes very sparing use of punishment or negative reinforcement, both an ethical bane to Skinner. The most frequent criticism of the novel was that the citizens who populated his fictional community were simply too happy. The novel has waxed and waned in its popularity over the decades, depending mostly on the whims of high-school literature teachers doing units on utopian societies. However, the book neither achieved a more widely held understanding of the operant basis of behavior nor inspired a popularized appreciation for Skinner's radical behaviorism among the larger readership he sought.

Skinner produced a range of other publications following *Walden Two* that were intended to challenge the prevailing wisdom. Particular note is warranted for his 1956 debate with Carl Rogers, the humanistic psychologist, at the annual meeting of the American Psychological Association. It provides a fascinating juxtaposition of humanistic and empirical approaches explaining human behavior; it is preserved for posterity in the journal *Science* (Rogers and Skinner, 1956) and in editions of his *Cumulative Record*. "Design of Cultures" (1961) is an additional noteworthy publication. The article is based on a series of three conferences on "Evolutionary Theory and Human Progress" held in 1960. Skinner offered an incisive analysis and criticism of scientific deference to Western philosophy when it comes to matters regarding scientific explanations of human behavior and culture. He openly challenged the prevailing custom to accuse science of "meddl(ing) in human affairs and infring(ing) on human freedoms," by presuming to take an empirical approach to the improvement of cultural practices.

Beyond Freedom and Dignity (1971) was Skinner's widely read and notable contribution addressing the "terrifying problems that face us in the world today." He quickly developed in his first chapter – "A Technology of Behavior" – a powerful argument to consider the efficacy of utilizing our best empirical data to inform the more fruitful construction of social systems to address the challenges of modern society. In Skinner's words,

> *The task of a scientific analysis is to explain how the behavior of a person as a physical system is related to the conditions under which the human species evolved and the conditions under which the individual lives . . . these events must be related . . .*
>
> *The effect of the environment on behavior remained obscure . . . We can see what organisms do to the world around them, as they take from it what they need and ward off its dangers, but it is much harder to see what the world does to them.*
>
> *The environment not only prods or lashes, it <u>selects</u>. Its role is similar to that in natural selection, though on a very different time scale, and was overlooked for the same reason. It is now clear that we must take into account what the environment does to an organism not only before but after it responds. Behavior is shaped and maintained by its consequences.*
>
> *By questioning the control exercised by autonomous man and demonstrating the control exercised by the environment, a science of behavior also seems to question dignity and worth. A person is responsible for his behavior, not only in the sense that he may be justly blamed or punished when he behaves badly, but also in the sense that he is to be given credit and admired for his achievements. A scientific analysis shifts the credit as well as the blame to the environment, and traditional practices can then no longer be justified. These are sweeping changes, and those who are committed to traditional theories and practices naturally resist them.*
>
> *Almost all our major problems involve human behavior, and they cannot be solved by physical and biological technology alone. What is needed is a technology of behavior.*
>
> *The environment is obviously important, but its role has remained obscure. It does not push or pull, it <u>selects</u> , and this function is difficult to discover and analyze. The role of natural selection in evolution was formulated only a little more than a hundred years ago, and the selective role of the environment in shaping and maintaining the behavior of the individual is only beginning to be recognized and studied.*

Freedom and dignity illustrate the difficulty. They are the possessions of the autonomous man of traditional theory, and they are essential to practices in which a person is held responsible for his conduct and given credit for his achievements. A scientific analysis shifts both the responsibility and the achievement to the environment. It also raises questions concerning the "values." Who will use a technology and to what end? Until these issues are resolved, a technology of behavior will continue to be rejected, and with it possibly the only way to solve our problems.

The foregoing quotations have been carefully selected from "A Technology of Behavior," Chapter 1 of *Beyond Freedom and Dignity*. His language and textual development became more precise during his decades of shaping the conceptual framework regarding freedom, dignity, and a technology of behavior, but Skinner had introduced these important considerations by the mid-1950s (see "Freedom and the Control of Men," reprinted in *Cumulative Record*). Empirical evidence consistently demonstrated the efficacy of his operant technology. What remained as Skinner's greatest challenge was the significant matter of cracking the Western philosophical resistance to applications of such technology in the more effective design of social systems. Illusions of free will, freedom, dignity, personal worth, and mind continue to be impediments to applications of those empirical advancements. The illusion of mind continues today as the most nagging problem in psychology. "There is no place in a scientific analysis of behavior for a mind or self" (Skinner, 1990).

Thus, it remained for Skinner to supplant his more philosophical approach to the problem of psychology with appeal to the larger community of scientists – particularly evolutionary biologists and ethologists. He produced two important publications in his last decade to that end. One appeared prior to his conversation with Wilson and the other, based on his 1990 presentation to the American Psychological Association, appeared shortly after his death. His 1987 request for a recorded conversation with Wilson was not by a coincidental accident; unquestionably, it was one more intentional effort to maintain the vitality of his mission and vision by thinking out loud with one of the most significant biologists of our time.

Selection by Consequences

There is no better way to identify this final section than to borrow the title Skinner used for his seminal 1981 article published in *Science*. Indeed, that article and several other Skinner publications were identified as "target" articles in a special edition of *The Behavioral and Brain Sciences*, published in the twilight of Skinner's career (7: 4, 1984). Consistent with the journal's format, the articles were accompanied by commentaries from many scholars that would potentially reinvigorate the flagging academic conversation regarding the magnitude and utility of Skinner's contributions. Commentaries were supportive, challenging, and occasionally laced with inevitable ideological tensions. However, Skinner's prominence continued to recede as work in cognitive psychology and the brain sciences quickly advanced into the last decade of the century.

"Selection by Consequences" is an article that is only rarely identified in contemporary scholarly literature and one to which the majority of psychologists and evolutionary biologists do not often make reference. It, along with "The Phylogeny and Ontogeny of Behavior," could likely be the most important products of Skinner's later years. They effectively capture the larger biological context for operant conditioning and place it squarely in the midst of evolutionary theory. Darwin's theory of evolution by natural selection and Wilson's sociobiology of cultural evolution are balanced by the intermediate selection process of Skinner's operant conditioning – the selection by consequence of behavior in the here and now. By way of summary, Skinner identifies that

> *human behavior is the joint product of (i) the contingencies of survival responsible for the natural selection of the species and (ii) the contingencies of reinforcement responsible for the repertoires acquired by its members, including (iii) the special contingencies maintained by an evolved social environment. (Ultimately, of course, it is all a matter of natural selection, since operant conditioning is an evolved process, of which cultural practices are special applications.)*

With one comprehensive stroke, Skinner aligned three of our most socially contentious theories regarding the ultimate nature of life and human nature. Skinner contends that three kinds of selection by consequence explain all aspects of physical and behavioral evolution. Three kinds of selection by consequence align evolutionary biology, behavioral psychology, and the special contingencies produced in the social environment necessary for gene-culture coevolution– sociobiology. Three kinds of selection by consequence connect the revolutionary contributions of Darwin, Skinner, and Wilson.

(1) Darwin's revolutionary theory of evolution describes how physical attributes (including the brain) were naturally selected on the basis of past ages-old consequences. These naturally selected attributes may or may not be relevant to the challenges of today's world. Time was large, and change was small and imperceptible in a lifetime. In reality, "it prepares a species only for a future that resembles the selecting past. Species behavior is effective only in a world that fairly closely resembles the world in which the species evolved." Such being the case, the relatively ultimate cause brain-based attributes selected were those providing long-term innate predispositionsto specific responses in the environment – sexual behavior, eating, fleeing, and defending territory. Yet, even these behavioral attributes may well be maladaptive in a world far removed from the African savannah on which they evolved. According to Skinner, this prospect for maladaptive behavior was corrected by the evolution of an absolute susceptibility for operantly emitted behavior to be reinforced (or punished) by their near-term consequences – operant conditioning.

(2) Operant conditioning provides the second type of selection by consequence. It evolved in parallel with an absolute susceptibility to the contingencies of reinforcement and a supply of potential behavior having little relation to causal stimuli. In fact, this free operant behavior oftentimes appears as volitional to naïve observers. Initially, Darwinian natural selection and operant conditioning worked together in a redundant fashion – operant conditioning providing reinforcement for appropriate

behavior that were initially guided by the innate tendencies previously determined by natural selection. Skinner frequently referred to the classic duckling experiment of Peterson (1960) to illustrate this point. The experiment demonstrated that the behavior of ducklings to follow their mother is both an evolved product of natural selection (imprinting) and an evolved susceptibility to reinforcement (protection, food, and contact comfort). Indeed, Skinner (1966) suggested that "what is inherited is not necessarily the behavior of following but a susceptibility to reinforcement by proximity to the mother or mother surrogate."

Eventually, "operant conditioning not only could supplement the natural selection of behavior, it could replace it." Its distinct advantage over evolution by natural selection and the evolution of culture is that operant conditioning is a kind of relatively ultimate cause real-time selection in progress. "It resembles a hundred million years of natural selection or a thousand years of the evolution of culture compressed into a very short period of time." Operant conditioning enables more behavioral flexibility and adaptability in the face of changing environmental demands – the relatively proximate cause changing environments in which innate and stereotypic responses could potentially be maladaptive.

When the evolution of susceptibility to the reinforcing strength of food and sexual contact emerged, new forms of behavior could be established. Elaborate customs of food gathering, preparation, and consumption appeared. Customs regarding sexual behavior that might eventually lead to sexual reinforcement were established that had nothing to do with the actual act of procreation. The reinforcing strength of these two survival needs shaped an ever-expanding array of behavior. The innate repertoires of behavior that begin to mimic social behavior are "within easy range of natural selection, because other members of the species are an ever-present feature of the environment." Innate social repertoires may then be supplemented by imitation, such as running when others run or smiling when others smile. This wider ranging imitation results from "the fact that contingencies of reinforcement which induce one organism to behave in a given way will often affect another organism when it behaves in the same way. An imitative repertoire which brings the imitator under the control of new contingencies is therefore acquired."

Skinner's long-time claim was that social behavior became more possible when the vocal musculature of humans came under operant control. Suffice it to say that "no new susceptibility to reinforcement was needed because the consequences of verbal behavior are distinguished only by the fact that they are mediated by other people," thereby greatly extending "the help one person receives from others. By producing verbal behavior people cooperate more successfully in common ventures. By taking advice, heeding warnings, following instructions, and observing rules, they profit from what others have already learned. The invention of the alphabet spread these advantages over great distances and periods of time."

(3) Modeling, priming, and the capacity for linguistic and its symbolic interactions are important aspects of Skinner's third kind of selection by consequences. The relatively proximate cause representational aspects of human behavior that show, tell, or teach (reify in Wilson's parlance) the appropriate behavioral repertoires all facilitate the emergence of social environments and culture. They also offer inter-

mediate (learned secondary) reinforcers to sustain the strength of the behavior until real-world reinforcers can take over. However, these intermediate reinforcers are usually available as an outcome of a group contingency and are quite powerful. That influence is a result of the enormous behavioral control exerted by special contingencies maintained in a social environment. Humankind is uniquely vulnerable to these special contingencies in that our capacity for wide-ranging verbal behavior

> *greatly increased the importance of a third kind of selection by consequences, the evolution of social environments or cultures. The process presumably begins at the level of the individual. A better way of making a tool, growing food, or teaching a child is reinforced by its consequence – the tool, the food, or a useful helper, respectively. A culture evolves when practices originating in this way contribute to the success of the practicing group in solving its problems. It is the effect on the group, not the reinforcing consequences for individual members, which is responsible for the evolution of culture.*

According to Skinner, the study of the evolution of culture is "primarily a matter of inferences from history." As such, it lacked the potential to be observed as an evolutionary process from beginning to end. He lamented that operant conditioning offered that observational capacity but had yet to achieve the widespread legitimacy it had earned in empirical demonstrations. He argued that "Sociobiology, for example, leaps from socio- to bio-, passing over the linking individual."

What Skinner failed to conceptually advance was the proposition that his mechanism of operant conditioning could be essential to describing the biological evolution of culture, thus providing the seamless linkage among the three kinds of selection by consequences. Additionally, his neglect of the vigorous discussions regarding multilevel selection theory and group selection that unfolded in the 1960s caused him to be inattentive to a serious consideration of how his conception of three kinds of selection by consequences might be aligned with multilevel selection and group selection. Darwin's natural selection explains the biological evolution of absolute susceptibility to reinforcing contingencies in the environment described by Skinner's operant conditioning. That real-time operant selection of individual behavior and the more significant longer-term operant selection maintained by the contingencies of the social environment may well be the relatively ultimate cause of gene-culture coevolution as described in E. O. Wilson's sociobiology. Skinner's inability to adequately resolve these concepts in no way diminishes the magnitude of his work; rather, time simply ran out for him.

Skinner sustained the legacy of his paradigm-shifting predecessors who pushed beyond prevailing wisdom and provided us new understanding of our human nature. He continued an intellectual tradition that nurtures empirical courage in the face of scholarly disregard and antagonism. He provided both inspiration and method for any scientist willing to venture into the controversial realm of human behavior and he flew in the face of his critics by self-describing his work as "radical behaviorism." In other words, Skinner admitted the controversial nature of his work by overtly acknowledging its "radical" nature. Skinner's name is another one added to the distinguished roll of scientists who challenged the status quo and advanced our understanding of human nature. At the same time, an inevitable outcome of his work was to make it even more apparent that humankind continued to recede as the center of the universe at an ever-increasing pace; not only was "mind" a useless construct, but humankind was deceiving itself to consider free will a reality. Skinner challenged our conceptions of freedom, free will, and the so-called dignity we earn by "freely" choosing to engage in altruistic behavior. Ultimately, he challenged our concept of self by claiming the contents of the "black box" were irrelevant.

Certainly, B.F. Skinner was among scholars who experienced challenges to his intellectual dignity during a career spanning the last half of the twentieth century. However, Skinner prevailed – most likely because of his vigorous intellectual commitment and his dogmatic insistence on empirically based objective explanations in the pursuit of wisdom. Skinner's legacy as the father of radical behaviorism and, perhaps, the most influential psychologist of the twentieth century is manifestly enlarged when we recognize how profoundly his foundational legacy continues. What remains is the challenge to recognize and understand the farther-ranging implications of his contributions and to enlarge the community of appreciation for his life work as it relates to a more integrated view of human behavior.

Chapter 2
Human Sociobiology: The Essential E.O. Wilson

> *What is human nature? It is not the genes, which prescribe it, or culture, its ultimate product. Rather, human nature is something else for which we have only begun to find ready expression. It is the epigenetic rules, the hereditary regularities of mental development that bias cultural evolution in one direction as opposed to another, and thus connect the genes to culture.*
> *(E.O. Wilson, 1998)*

Paradigm shifts are refreshing new ways for humanity to understand the nature of their existence and their universe. Sadly, any challenge to the commonly held current order promises to introduce significant dissonance to most practitioners in a discipline. Those responses are no less dramatic in empirical practice than in the practice of literature, economics, politics, and the wide array of other disciplines. Thus, the more frequent response to the introduction of a new paradigm is disbelief, perceived threat, antagonism, and outright attack. This common occurrence was elegantly noted by Darwin in the conclusion of *The Origin of Species* (1859) when he acknowledged "... I by no means expect to convince experienced naturalists whose minds are stocked with a multitude of facts all viewed, during a long course of years, from a point of view directly opposite to mine". Surely, as we recognize the sesquicentennary year of Darwin's conception of biological evolution, we must acknowledge that it competes in most minds with all other creation myths (Durant, 1980).

Paradigm shifts can also be ideological threats as much as they are threats to a prevailing world view, causing irrational emotional responses couched as intellectual challenges. Such responses are based on tenaciously held personal ideologies that emerge over a lifetime of personal experience and emotional investment. Frequently, they are reinforced by the real-life need to sustain intellectual stature and professional influence. Most regrettable are those challenges roused by political or religious ideologies that have nothing at all to do with observations that are at hand, but have everything to do with the maintenance of a personal world view. Undeniably, true paradigm shifts are often challenged with such descriptors as controversial, untested, irresponsible, out of line with prevailing wisdom, or inconsistent with the evidence. Occasionally, the new paradigm and its proponents are publicly

P. Naour, *E.O. Wilson and B.F. Skinner*, Developments in Primatology: Progress and Prospects, DOI 10.1007/978-0-387-89462-1_2,

castigated with inflammatory rhetoric – such as blinded by bias, racist, or sexist. E.O. Wilson was not immune from such angst. Without a doubt, his 1971 naming and establishment of sociobiology as a discipline (*The Insect Societies*) both quickened the pace of humankind's anthropocentric demise and accelerated the perceived controversial nature of his work. Regardless of the manifestly important groundwork produced by scholars who articulated the foundations of sociobiology, it was Wilson himself who became the embodiment for its public persona.

For the first time, a comprehensive paradigm, a "new synthesis," was offered that no longer treated human social behavior and biology as separate subjects. Simply stated, sociobiology was defined by Wilson as "the systematic study of the biological basis of all forms of social behavior." By extension, the evolution of social behavior and culture is driven by the environmental contingencies of natural selection – just like any other behavioral repertoire or physical attribute. To be sure, *all* behavior should necessarily be viewed as an extension of the brain as a biological entity – including the repertoire of behavior typically described as human culture. The early 1970s witnessed a growing number of scholars ready to take on the rapidly accelerating sociobiology debate that was further animated by Wilson's *Sociobiology: The New Synthesis* (1975), and other publications such as *Animal Behavior: An Evolutionary Approach* (Alcock, 1975), *The Evolution of Behavior* (Brown, 1975), *Ethology: The Biology of Behavior* (Eibl-Eibesfeldt, 1975), and *The Selfish Gene* (Dawkins, 1976).

Predictably, this growing cadre of sociobiologists – Wilson in particular – was not immune to intellectual antagonism and verbal abuse related to sociobiology generally and human sociobiology in particular. Some of the most rancorous commentary was provoked by the so-called Sociobiology Study Group, the creation of scientist ideologues Stephen Jay Gould and Richard Lewontin. Their Marxist critique of Wilson went so far as to accuse sociobiology of being friendly to racism, misogyny, sexism, and genocide – inspiring verbal harassment of Wilson at public appearances and a well-known ice water dousing at the 1978 gathering of the American Association for the Advancement of Science. Readers interested in an authoritative discussion of the debate, its key players, and their work are encouraged to refer to Ullica Segerstråle's comprehensive and superbly crafted *Defenders of the Truth* (2000). To say the book is comprehensive somehow diminishes the magnitude of its accomplishment in meticulously capturing an essential moment in the history of science.

Wilson weathered the storm of criticism and attack with his characteristic kindness and dignity, personal attributes that have endeared him to the larger intellectual community during his long and productive career. Perhaps wisely, he retreated from the sociobiology debate during the decade of the 1980s, following the publication of *Genes, Mind and Culture* (1981) to focus more of his intellectual energy on his scientific passion for ants, environmental sustainability, and his newly developing concept of biophilia – all resulting in significant publications during that time. That passage of time dulled the edge of the ideological blade brandished at Wilson. Yet, the occasional distortion (both unintended and intended) and misrepresentation of his work persist to this day – not unlike the experience of other paradigm challenging scholars that preceded him. Now, more than three decades after the initial publication of *The Insect Societies* and *Sociobiology: The New Synthesis*, Wilson's work is more widely accepted and is rightfully recognized as the essential foundation for the rapidly expanding field of evolutionary psychology.

Scholars have more recently offered Wilson a far more balanced and less biased reading, although the sociobiology conversation is far from over and its theoretical foundations far from commonly established among its scholars (Wilson and Wilson, 2007). Certainly, since 2004 (Wilson and Hölldobler), he has moved experts in genetic social theory of nonhuman animals to discuss (and perhaps rescue) group selection as a key element of sociobiology. Notwithstanding these important theoretical conversations that continue, he has rightfully earned a prominent place in the larger history of science and biology. Wilson deserves an open-minded readership for and balanced evaluation of his important ideas. Thus, the major portion of this chapter provides an overview of the essential conceptual framework of Wilson's sociobiology framed in the larger scientific and historical context of Darwinian evolutionary theory.

The HMS Beagle to the Modern Synthesis

Wilson's 1975 "new synthesis" challenge was clearly an innovative and effective means to enlarge the conversation regarding the fundamental biological elements of human nature and engage the social sciences in that larger intellectual conversation. Similar to any work of this intellectual magnitude, it did not come quickly or in isolation – many great minds set the stage. First and most notable among those great minds was Charles Darwin, on whose considerable shoulders Wilson stood in order to view the new horizon of sociobiology. Thus, a description of Darwin's theory of evolution by natural selection is the necessary starting point, followed by an account of other key contributors in the advancement of evolutionary theory in the early twentieth century, its more recent modern synthesis, and Wilson's new synthesis.

Darwin's predilection to keenly observe the natural world is mirrored in the early life of Wilson. Darwin was naturally drawn to beetles as a child, much like Wilson was drawn to ants. Innate and wide-ranging curiosity, the naturalist's penchant for keen observation, compulsively meticulous cataloging, and the precious ingredient of time were common elements in their early lives. Indeed, it is no irony to those familiar with their early lives and the scientific importance of their work as adults that Darwin and Wilson had so much in common. The trajectory of their intellectual lives was clearly established in their youth. A similar alignment is common to their adult lives, in that a keen focus on specific organisms at any given point in their career was complimented by enormous and unbounded appetite for all elements of the natural world. In Darwin's case, that appetite took him from barnacles to human morality – for Wilson, from ants to great apes and human culture.

Most biographers of Darwin identify his personal conflict in determining a professional direction for his early adult life. Medical practice and the ministry were both eliminated very quickly by Darwin as appropriate professional pursuits in favor of natural history and beetle collection. Instead, his lifelong curiosity for the natural world inspired his more active pursuit of botany and geology. His formal study of both disciplines was undertaken in a timely enough fashion for his achievement to gain notice as a prospect to serve as the unpaid naturalist for the voyage of the HMS Beagle in 1831. The 23-year old Darwin was offered that opportunity and he set sail on a 5-year voyage that originated to complete a detailed mapping of the South American coastline. The voyage included lengthy and plentiful stops along the way, enabling Darwin to build an enormous specimen collection. The voyage's circumnavigation of the globe before its return to England provided Darwin enormous time to carefully examine, describe, and catalog the specimens.

Darwin's interest in geology moved him to include a copy of Charles Lyell's recently published *Principles of Geology* (1830) on his Beagle journey. Lyell's book firmly established for Darwin two essential concepts to advance his thinking – the dynamic nature of the earth's physical history and its immense age. The trip provided ample additional evidence to support those concepts. While ashore on the west coast of South America, Darwin made particular note of the fossilized remains of a long extinct South American giant ground sloth and personally witnessed the dynamic earth at work when he experienced an earthquake. Additionally, the voyage provided enormous evidence for the diversity of life forms on the planet and very direct experiences with several human cultures that differed from those to which he was accustomed on the European continent. He surely achieved some of his most elegant writing as he deconstructed empty criticisms regarding perceived gaps in the fossil record. At the same time he provided example after example of observed fossilized extinct species and offered magnificent descriptions of transitional species in the time dependent process of evolution by natural selection (see *The Origin of Species*, Chapter 10). Darwin's genius for detailed analytical observation and his remarkable capacity to synthesize across those observations gave life to a new vision for the natural history of the planet.

The last spark of inspiration was provided by his fortuitous reading of a lengthy essay by Thomas Malthus, *An Essay on Population* (1798), two years after the return of the Beagle (Darwin, 1929). Malthus, an English economist, offered a theory to explain the occurrence of famine in populations. We now know his mathematical assumptions to be inaccurate; however, his theory that famine would inevitably result from the pressures of population growth was correct and an essential component in Darwin's thinking. Malthus conjectured that food resources tended to increase in only arithmetic progression, while populations competing for those resources increased in geometric progression. Overpopulation results in famine. Darwin pushed that thinking to suggest that the natural limitation of food resources might more logically result in individual competition for those resources rather than cause famine. Those organisms succeeding in that competition lived to achieve reproductive age and likely would pass to their offspring the physical and behavioral characteristics that resulted in their success.

> *I soon perceived that selection was the keystone of man's success in making useful races of animals and plants. But how selection could be applied to organisms living in a state of nature remained for some time a mystery to me.*
>
> *...I happened to read for amusement Malthus on* Population, *and being well prepared to appreciate the struggle for existence which everywhere goes on from long-continued observation of the habits of animals and plants, it at once struck me that under these circumstances favourable variations would tend to be preserved, and unfavourable ones to be destroyed. The result of this would be the formation of new species. Here, then, I had at last got a theory by which to work.*
>
> (*Darwin, 1929*)

The serendipitous convergence of the voyage, well chosen books, and natural curiosity during the formative years of Darwin's scientific apprenticeship established the foundation for his life's work. The coincidental alignment of those factors during an era in which science and empirical practice were increasingly held in high regard provided fertile ground for the seeds of his revolutionary insights to grow. Observation and quantification, the hallmarks of empiricism, became the acknowledged essentials of scientific best practice – Darwin was obsessive about those basic empirical attributes. These elements combined to result in the most important biological statement of modern time – the 1859 publication of *The Origin of Species*, detailing Darwin's theory of "descent with modification through variation and natural selection."

His detailed observations made it apparent that individuals within a species have subtle differences in both physical and behavioral characteristics. It was also obvious to him that species have changed over an immensely long natural history of the earth. That gradual change was characterized by Darwin as evolution driven by the mechanism of natural selection. In sequential summary form, the following observations are commonly accepted to be those which led to Darwin's conceptualization:

(1) The age of the planet is immensely greater than previously accepted by theologians and academics. The magnitude of that time is on the order of millions of years during which the natural physical history of the earth has been fluid (Lyell).
(2) Population growth inevitably outstrips the resources necessary to sustain all individuals and results in famine (Malthus) or competition for survival (Darwin).
(3) Individual members within a species have observable physical and behavioral differences (Darwin).
(4) Individuals with differences that enable them to more successfully compete for limited resources will more likely survive to reproductive age. Consequently, it becomes likely they will pass on those characteristics to their offspring (Darwin).
(5) Darwin described this competition as natural selection – the process whereby small, but favorable, differences are passed on to the offspring and ultimately achieve common expression in future generations. The gradual accumulation of those differences can result in the emergence of new species and the extinction of an existing species.

A close reading of Darwin reveals his uncanny capacity to intersperse his lengthy and dense objective descriptions of the natural record with rich commentary that approaches the majesty of poetic achievement. Some of his best prose is accomplished in the book's concluding paragraphs in which he invites his readers to step back from the overwhelming detail of the book and view the magnificent simplicity of his theory.

> *It is interesting to contemplate a tangled bank, clothed with many plants of many kinds, with birds singing on the bushes, with various insects flitting about, and with worms crawling through the damp earth, and to reflect that these elaborately constructed forms, so different from each other, and dependent upon each other in so complex a manner, have all been produced by laws acting around us ... Thus, from the war of nature, from famine and death, the most exalted object which we are capable of conceiving, namely, the production of higher animals, directly follows. There is grandeur in this view of life.*
>
> *(Darwin, 1859)*

Darwin's theory in reality is not limited to just natural selection. Rather, it includes five related theories, each of which received varying levels of acceptance and challenge by late nineteenth and early twentieth century scholars. Simply stated, the five theories developed by Darwin include

(1) the ever-changing nature of species – nonconstancy made possible by genetic recombination of sexual reproduction and gene mutation,
(2) all organisms have common ancestry – organismal diversity "branches" by its nature,
(3) evolution is gradual – there are no biological discontinuities, only discontinuities in the fossil record that demonstrate gradual emergence of species,
(4) the slow but constant emergence of new species that lead to diversity and,
(5) natural selection.

The common pursuit of many scientists for many years following the publication of the book was to argue the reality and then the utility of these individual concepts, rather than to see them relating to each other by necessity. Still today, it is quite common that natural selection is exclusively identified by most as Darwin's chief contribution. Yet, Darwin was explicit from the outset in describing the essential interrelationship among all five of the concepts as the foundation of the origin of species. His critics tended to pull the concepts out of context from each other and thereby diminish Darwin's work by misrepresenting its internal consistency and, ultimately, the beauty of its simplicity.

Not until the 1920s, with the emergence of neo-Darwinism, did scholarly work begin to achieve a recognizable synthesis among the five theories with the newly developing field of population genetics. More significant theoretical foundations for accomplishing the synthesis were established in the 1930s, "When the publication of Ronald Fisher's *The Genetical Theory of Natural Selection* (1930), Sewall Wright's *Evolution in Mendelian Populations* (1931), and J.B.S. Haldane's *The Causes of Evolution* (1932)" demonstrated increasing acceptance among scientists – population geneticists in particular – for the interrelated nature of the five theories (Wilson,

1975). Ultimately, Theodosius Dobzhansky's *Genetics and the Origin of Species* (1937) became widely recognized as the fiber uniting Darwin's theory with modern genetics. Dobzhansky's synthesis had immense influence on a generation of biologists, and soon came to be recognized as the next major advancement in evolutionary theory – the so-called Modern Synthesis (Mayr, 1942). The fabric representing the modern era was then woven by Dobzhansky's contemporaries, including Julian Huxley (general theory), G. Evelyn Hutchinson (ecology), Ernst Mayr (systematics), G.G. Simpson (paleontology), G. Ledyard Stebbins (botany), and M.J.D. White (cytology). The full synthesis of their work in the larger context of Dobzhansky resolved the disconnection of various intellectual approaches to evolution and provided direction for modern evolutionary thought.

The Modern Synthesis made it apparent that the contemporary basis for understanding evolution would be described by the natural selection of variations caused by small genetic changes and that the more significant changes we observe in the fossil record can be explained by these same mechanisms continuing over enormous periods of time. Only in the very recent past (reviewed in Carroll, 2005 and Amundson, 2005) has a significantly new approach to evolutionary investigation emerged, albeit outside the framework of the modern synthesis. Evolutionary developmental biology (popularly known as evo devo) opened that fresh avenue of evolutionary investigation by describing mechanisms that go beyond natural selection and the gradual accumulation in populations of small genetic variations as the bases for evolution. Instead, evo devo explores the genetic and developmental mechanisms that guide how the actual form of organisms has evolved. Researchers are investigating how organismal morphology can be modified by evolutionary alterations in the ontogenetic process and how subtle changes in ontogeny produced by "toolkit genes" or "genetic switches" can alter individual development enough to initiate speciation.

Sociobiology and the New Synthesis

Wilson (1975) characterized the Modern Synthesis as "the elucidation, through excellent empirical research, of the nature of genetic variation within species and of the means by which species multiply." This synthesis was made possible when the newly emerging "branches of evolutionary biology – systematics, comparative morphology and physiology, paleontology, cytogenetics, and ethology were reformulated in the language of early population genetics." Stated simply, the Modern Synthesis weighs "each phenomenon for its adaptive significance and then relates it to the basic principles of population genetics." Sadly, until the recent emergence of evolutionary psychology and evolutionary developmental psychology, the social sciences made only slow and oftentimes contentious progress toward inclusion in the Modern Synthesis. Wilson's 1975 publication of *Sociobiology: The New Synthesis* intentionally set out to challenge this record by expanding the synthesis beyond the domain of biology to include the social sciences; thus, the basis of his subtitle –

The New Synthesis. Ultimately, Wilson aspired to "reformulate the foundations of the social sciences in a way that draws them into the Modern Synthesis," and he is meticulous in systematically developing the case that sociobiology completes the Modern Synthesis by drawing social science into evolutionary biology. His more recent *Consilience* (1998) is even more eloquent and expansive in describing that "new synthesis," while challenging scholars to advance the new synthesis to include the humanistic social sciences and humanities.

Scholars have slowly recognized that one of Wilson's most important and lasting intellectual contributions has been to make that Modern Synthesis real. Inspired by the central dogma of evolutionary biology that natural selection shapes ALL classes of traits in organisms, Wilson drew upon the work of ethologists, like Konrad Lorenz, to theorize that behavior and social structure should be studied as "'organs,' extensions of genes that exist because of their superior adaptive value." Thus, Wilson's essential definition of sociobiology describes it as "the systematic study of the biological basis of all forms of social behavior" (Wilson, 1975).

Wilson earned significant regard as a scientist early in his career. He established himself as one of the world's foremost experts on ants and their behavior in the 1950s, publishing extensively throughout that decade. His work on chemical communication among ants emerged in the late 1950s and guided the focus of his interest toward the biological basis for social structures among ants and beyond. One major result of that era in Wilson's intellectual life was his benchmark publication, *The Insect Societies* (1971). Among its many important conceptual contributions was to establish and name sociobiology as a discipline. To this day, the contents of the premier journal *Behavioral Ecology and Sociobiology* are built upon the original formulation that the discipline of sociobiology would be established on the foundation of population biology – a relatively new discipline in itself in the 1960s and 1970s. Thus, "sociobiology addresses not just the genetic origin of eusociality, but all of the key phenomena in communication, division of labor, and colony life cycles" (Wilson, 2007).

Wilson's early interests led him to consider the wider-ranging implications of social behavior, biology, and evolution. His greatest attention was on insect societies – the organization of their populations, caste structure, mechanisms for communication, and the underlying physiology of social adaptations. Yet, at the same time, he methodically pushed his conceptual framework to include all animal societies. Those more comprehensive considerations became the basis for his 1975 publication of *Sociobiology*, launching him into a much larger arena of attention and controversy. *Sociobiology* is most significantly dedicated to a comprehensive cataloging of the biological basis of social behavior across the great chain of being and succeeds in achieving a comparative view. Sparingly, he pushed the theoretical envelope to include primates and, finally, humankind in the final chapter. Yet, it is not by necessity of inclusion in that great chain that humankind appears in the new synthesis; it is by design. From the outset, Wilson intended for *Sociobiology* to have an ultimate focus on the social behavior of humankind. He does so by initiating his discussion of human sociobiology with an explication of the adaptive features of organization in small and isolated contemporary human societies. By extension, we are all included. Yet, it is not until the 1981 publication of *Genes, Mind and Culture* that Wilson offers the mechanism for human sociobiology, gene-culture coevolution.

These initial comments are not intended to suggest that Wilson's work emerged in isolation or appeared as the only scholarly work advancing sociobiology. In fact, most scholars point to William Hamilton's publications of the early 1960s as the conceptual birth of sociobiology. Hamilton suggested that it was kin selection, rather than group selection, which explained the evolution of altruistic behavior. His early work on a genetic theory of social behavior and the evolution of altruistic behavior are commonly regarded as essential to the emergence of sociobiology. Indeed, Wilson credits Hamilton's theory of kin selection (1964) for providing him the key conceptual spark to invigorate his theoretical development of sociobiology. Yet, Hamilton's work also prompted the rejection of group selection as a mechanism driving the evolution of altruistic behavior. The rejection achieved such prominent stature within the community of evolutionary biologists that group selection was consistently avoided in evolutionary theory and sociobiology as a topic relevant to its theoretical foundations (reviewed in Williams, 1966). Additionally, scientifically naïve accusations regarding the determinism of group selection had significant enough political overtones to convince most scholars to maintain an assured clear distance from the concept.

Hamilton's work initiated more focused attention on a possible selfish gene mechanism operating in kin selection. Mathematically oriented theorists, such as George Williams, Robert Trivers, and John Maynard Smith (Evolutionarily Stable Strategy), all continued the march toward sociobiology in important ways. Most prominently, the early and continuing contributions of Richard Dawkins (*The Selfish Gene*, 1976; *The Extended Phenotype*, 1982; and *The Blind Watchmaker*, 1987) continue to be viewed by many as essential reading for a comprehensive understanding of the twentieth century sociobiology conversation. (Interested readers are also encouraged to read Grafen's and Ridley's book of collected essays, *Richard Dawkins: How a Scientist Changed the Way We Think*, for a more detailed analysis.) However, the long-lived impulse to avoid group selection in favor of kin selection and the selfish gene led to theoretical pursuit based solely on individual and genetic self-interest. This avoidance became so much a part of standard operating procedure in evolutionary biology that a generation of scholars was trained with no foundation in group selection as a component of multilevel selection (Wilson and Sober, 1994).

The most recent twenty-first century rethinking of the theoretical foundation of sociobiology holds the potential to correct the bearing of contemporary theory in sociobiology. E.O. Wilson now maintains that kin selection was an element, but not the principal founding idea, of sociobiology and that "kin selection theory in its original form has collapsed" (Wilson, 2007). Further, Wilson and Wilson (2007) suggest that current sociobiology is in "theoretical disarray," since the majority of scholars are still reluctant to revisit the 1960s rejection of group selection. Consequently, the development of alternative explanations for the evolution of cooperative and altruistic behavior continues to be sorely hampered. This current disarray has stirred a vigorous conversation among a growing number of sociobiologists and has resulted in an important recapitulation of multilevel selection theory, including group selection, as the theoretical foundation for sociobiology in the future. "The importance of group selection in human evolution enables our groupish nature to be explained at face value. Thus, multilevel selection, not group selection alone, provides a comprehensive framework for understanding human evolution along with other major transitions." This "new" theoretical foundation of sociobiology takes us full circle to Darwin's suggestion that "natural selection takes place on more than one level of the biological hierarchy. Selfish individuals might outcompete altruists within groups, but altruistic groups outcompete selfish groups. This is the essential logic of what has become known as multiple selection theory" (Wilson and Wilson, 2007):

> *Like an exploded diagram of a machine, it (multilevel selection) allows one to identify the component parts of evolution in metapopulations and to see how they fit together. Natural selection within groups, variation among groups, and the way in which groups contribute to the formation of new groups are all fairly easy to understand as separate processes, after which they can be put together to determine what evolves.*
>
> *(D.S. Wilson, 1998)*

Human Sociobiology

> *The consideration of these various facts impresses the mind almost in the same manner as does the vain endeavor to grapple with the idea of eternity.*
>
> *(Charles Darwin, 1859)*

The fossil record produced over millions of years reveals the slow march of natural selection in response to the ever-changing physical environment and supports as fact the five theories of Darwinian evolution. Even the most capable scientific minds in Darwin's era and after had enormous difficulty resolving the magnitude of time necessary for that slow march to proceed to no particular destination. Time and the suspension of perceived direction or progress as components of evolution were, and continue today as, significant obstacles to understanding evolution by natural selection. Additionally, humankind's common tendency toward anthropocentrism and need to view humanity as the crowning achievement of life on planet earth continue as barriers to better scientific understanding of evolution by natural selection. To this day, many prominent scientists deny the sociobiological essence of human culture and insist that promethean genes effectively liberate humankind from the remainder of human DNA and somehow free the mind from all antecedent biology. (Lumsden and Wilson, 1983)

Anthropocentrism achieves an even greater significance and potential for controversy when the evolution of human behavioral repertoires and constructs like culture are considered in light of evolutionary theory. This was the intellectual context in which E.O. Wilson's *Sociobiology: The New Synthesis* (1975) was published, little more than a century after the appearance of *The Origin of Species* (1859). Equal in its scientific rigor and comparable in its written elegance, *Sociobiology* initially roused similar antagonism and received only limited acceptance. Its twentieth century Darwinism further challenges the place and importance of individual existence by suggesting, "... in evolutionary time, the individual organism counts for almost nothing ... Its primary function is not even to reproduce other organisms; it reproduces genes, and serves as their temporary carrier."

Human sociobiology brings about its most significant challenge to prevailing wisdom by suggesting that naturally selected genes provide the blueprint for the ENTIRE organism – including the brain, which carries predispositions for ALL behavioral repertoires. The leap from the natural selection of physical attributes to behavioral predisposition and expression of social behavior is the tension. Wilson advanced the concept that the emergence of culture is ultimately an expression of the natural selection of genes coding for those behavioral predispositions, natural selection being a "process whereby certain genes gain representation in the following generations superior to that of other genes located at the same chromosome positions." The environment and experience will then act to shape the trajectory of those genetic predispositions (Wilson, 1975).

Wilson went on to make human sociobiology more accessible in his Pulitzer Prize winning 1978 publication, *On Human Nature*. The question that has defined the most significant intellectual contributions of Wilson's long and prolific career is identified at the outset of *On Human Nature*; "How does the mind work, and beyond that why does it work in such a way and not another, and from these two considerations together, what is man's ultimate nature?" Along with Charles Lumsden, Wilson also developed sophisticated mathematical modeling for gene culture coevolution in the 1981 book, *Genes, Mind and Culture* – a book that is infrequently referenced, but nonetheless essential to more completely understanding the evolution of human social behavior and culture. They also jointly authored a more accessible version of this 1981 publication, with *Promethean Fire* in 1983. In its own way, each subsequent publication reinforced human sociobiology's focus on mind and humankind's ultimate nature.

Wilson's final chapter of *Sociobiology* asserts his answer to the question of humankind's ultimate nature, "... the biological principles which now appear to be working reasonably well for animals in general can be extended profitably to the social sciences." His sociobiology, as the "new synthesis" of biology and the social sciences was conceptualized to achieve a new naturalism. "But to the extent that new naturalism is true, its pursuit seems certain to generate two great spiritual dilemmas" (Wilson, 1978). The two dilemmas are at the core of *human* sociobiology. Humankind's capacity to resolve the ideological and emotional conflict that results from acknowledging the two dilemmas, and to objectively describe their mechanisms will determine the prospects for the long-term success of our species.

Dilemma 1: "No species, ours included, possesses a purpose beyond the imperatives created by genetic history. Species may have vast potential for material and mental progress but they lack any immanent purpose ... the human mind is constructed in a way that locks it inside this fundamental constraint and forces it to make choices with a purely biological instrument. The human mind is a device for survival and reproduction, and reason is just one of its various techniques." The implication of this dilemma is that transcendent and defining societal goals quickly diminish in their presumed moral equivalency of war. "In order to search for a new morality based upon a more truthful definition of man, it is necessary to look inward, to dissect the machinery of the mind and to retrace its evolutionary history." Therein lies the basis for the second dilemma.

Dilemma 2: "Innate censors and motivators exist in the brain that deeply and unconsciously affect our ethical premises; from these roots, morality evolved as instinct. Human emotional responses and the more general ethical practices based on them have been programmed to a substantial degree by natural selection over thousands of generations. . .Which of the censors and motivators should be obeyed and which ones might better be curtailed or sublimated? These guides are the very core of our humanity."

Reductionism is explicit in these dilemmas – a reductionism that is essential to the empirical process, yet which is frequently rejected by many academics when applied to humankind and our behavior. Even many of our most accomplished scientists have not resolved their emotional tension regarding the prospect that we are dehumanized by describing our behavior consistent with the laws of biology. "This perception, which equates the method of reduction with the philosophy of diminution, is entirely in error. The proper study of man is, for reasons that now transcend anthropocentrism, man" (Wilson, 1978). Thus, Wilson has advanced sociobiology as the systematic study of the biological basis of all forms of social behavior, in all kinds of organisms, including man. More specifically, sociobiology is ". . . a hybrid discipline derived from ethology, ecology, and genetics in order to derive general principles concerning the biological properties of entire societies" (Lumsden and Wilson, 1981). Sociobiology has accomplished this by reassembling the primary characteristics of social organization from ecology and genetics studied at the population level to show how social groups adapt to the environment by evolution. Ultimately, Wilson seeks application of sociobiology to better understand how we might shape human culture to increase altruism and cooperation in favor of our behavioral predispositions such as aggression and xenophobia. Such self-destructive behavior are relics of the primitive selection conditions that shaped behavior in our evolutionary past.

Wilson advanced a more complete conceptualization of the evolutionary basis for human sociobiology along with Charles Lumsden in the 1981 publication of *Genes, Mind and Culture*. The book offered an in-depth theoretical and mathematical framework for the mechanisms that "connect individual mental development to culture and culture to genetic evolution." With the support of Lumsden, a physicist who extended his research interests into theoretical biology, the book also provided explicit mathematical models for those mechanisms. They suggested that the overarching concept driving the genetic basis of social behavior and culture is *gene-culture coevolution*, a still largely underappreciated concept. They described it as a "complicated, fascinating interaction in which culture is generated and shaped by biological imperatives while biological traits are simultaneously altered by genetic evolution in response to cultural innovation." Sadly, few scholars have given much attention to gene-culture coevolution as a more comprehensive theory than Dawkins' selfish genes and memes to explain human sociobiology (see Chapter 3 for a brief description of meme theory).

Gene-Culture Coevolution

The years since the appearance of *Genes, Mind and Culture* have been witness to an increasing capacity of social scientists to see beyond their initial rejection of biological causation and natural selection as the drivers of culture. Those years have seen the steady fruition and more widespread acceptance of evolutionary psychology. Lumsden and Wilson suggest that some initial reluctance among scientists might have been generally based on what some called the promethean gene hypothesis, which posits that a group of genes essentially freed the human mind from other genes by their function to provide for only the *capacity* to evolve by culture – not to provide for the kind of rich interactive feedback process described by gene-culture coevolution. *Promethean Fire* (1983) is their attempt to argue that the development of the human mind is not somehow freed from the biological imperatives of gene-culture coevolution. Instead, the seamless fabric of evolution is all-inclusive and needs no special stitching to hold the emergence of human nature and the human mind in place.

More recently, evolutionary psychology has established a more integrated direction for the social sciences toward alignment with the principles of biological evolution. Thus, human social behavior and culture are increasingly viewed by the social sciences as evolving and dynamic biological processes. David Buller (2005) offers a critical examination of evolutionary psychology and its chief contributors in *Adapting Minds*; readers are encouraged to seek out his appraisal of the discipline for a contemporary perspective. The field has generally gotten beyond the early criticisms of human sociobiology as prescriptive and deterministic – criticisms that were far more ideological than rational. Instead, there is a more sophisticated understanding that social behavior and culture are not explicitly prescribed in the genes; rather, they are implied by gene-culture coevolution. In other words, genes coevolve with culture to "prescribe a set of biological processes, called *epigenetic rules* that direct the assembly of the mind" (Lumsden and Wilson, 1981).

Epigenetic rules channel development of individual behavior according to the predispositions coded by the gene ensembles inherited by an organism. Ultimately genetically based, they are a set of biological processes that are both gene dependent and context dependent – context provided by information "derived from culture and physical environment." The human species is not immune to those epigenetic rules. We are as much a product of the coevolutionary processes as all primate species, "each adapted in idiosyncratic ways to particular environments." These paradigm challenging concepts require unfailing vigilance to allay the temptation by critics to irrationally attack their scientific basis. In Wilson's own words, "behavior is not explicit in the genes, and mind cannot be treated as a mere replica of behavioral traits." Rather, "genes prescribe a *set of biological processes* (epigenetic rules) that direct the assembly of mind." Culture emerges as the collective behavior of many individuals aligning to create cultural patterns. Yet, the behavior of each individual still determines survival and prospects for reproduction – thus, their genetic fitness and the rate at which the gene ensembles dominate or decline within the population.

Epigenetic rules describe a mind as a system that tends to organize into certain forms in preference over others, while the combined action of many minds seems to lead to the emergence of patterns in culture that become statistically predictable. Culture, then, may be described as the mass pattern of mental activity, mental constructs, and behavior emitted by each organism as a result of expressed epigenetic rules. It is the process of social learning that enables the transmission of culture from one generation to another. Mind is the construct that expresses those cultural attributes within the individual organism. In the view of Lumsden and Wilson, genes are linked to culture and the individual construct of mind in a very subtle fashion. That gene-culture linkage achieves so consistent a path in its result across humankind that their biological foundation is a logical conclusion.

The biology of cultural transmission that achieves its ultimate result in the emergence of the human mind is driven by the behavioral attributes of learning and teaching. Lumsden and Wilson specify four capacities related to learning and teaching that appear in order of increasing sophistication: (1) learning – refers to classical conditioning; (2) imitation – defined by simple mimicking; (3) teaching – includes those operantly conditioned behaviors that are often initiated by cuing and (4) reification – the graphic or verbal symbolic representation of conceptual information. These four are the behavioral capacities organisms have evolved that are at the heart of cultural transmission. They can be combined in a variety of ways as means to structure stimuli, environments, and contingencies to channel behavioral response in organisms and, ultimately, to achieve symbolic conceptual representation in the human mind. Most typically, the attributes also appear in this increasing order in most lines of animal evolution. Lumsden and Wilson offer a simple combination of the components to describe five evolutionary grades that might parallel the emergence of cultural behavior. They propose that each species fits within one of the evolutionary grades and that the number of species falls rapidly at each subsequent grade. They suggest that humans alone occupy the final eucultural state defining true culture.

	Learning	*Imitation*	*Teaching*	*Reification*
Acultural I				
Acultural II	X			
Protocultural I	X	X		
Protocultural II	X	X	X	
Eucultural	X	X	X	X

Learning – Respondent conditioning
Imitation – Cue and mimic
Teaching – Cue/mimic and operantly shape
Reification – Symbolic representation what we desire to transmit intergenerationally (words, etc.)

(from: Lumsden and Wilson, 1981. p. 3)

Culture is intergenerationally transmitted by all four attributes. However, reification alone provides the clear separation of humankind from other organisms as it achieves the eucultural state. We achieve that capacity by means of highly structured learning environments (school) that promote specific stages of cognitive development ultimately resulting in the emergence of mental operations enabling sophisticated symbolic representation of abstract conceptual information – the stuff of Wilson's culturgens (memes). Reification (symbolization) assumes the capacity to produce those abstract concepts and continuously reclassify the world in the face of ever-increasing accumulated experiences and "creates order in a world otherwise overwhelmed in flux and detail" (Wilson, 1998).

The developmental nature of those mental operations was the basis of immense research over the last half of the twentieth century – most embodied in the life work of Jean Piaget. The scholarship of Piagetian developmental psychology was important in guiding our understanding that the mental operations required of abstract representation and reification come only after each child moves through a very specific sequence of cognitive development. Each child must first move through (1) a sensorimotor stage limited exclusively to here and now sensory experience, (2) a preoperational stage in which language is first used but significantly limited by both preconceptual and intuitive mistakes that result from inadequate experience and incomplete language, (3) a concrete operations stage in which language is increasingly utilized as a tool of mental representation and classification, but is also limited by concrete representation and category inflexibility, and (4) a formal operations stage that is characterized by reification and our adult capacities for abstract symbolic conceptual representation.

Reification enables the human mind to classify and represent sensory experience, it is a means to represent memories, it associates emotional context to memories and triggers emotional responses, and it is the means by which we transmit information and feeling to others. Lumsden and Wilson (1981) submitted that the attainment of euculture as a result of humankind's increasing capacity to reify was a unique event in the evolution of our species and the emergence of mind. "It was achieved through an acceleration of neuroanatomical and behavioral evolution unprecedented in the history of life. One can visualize the process in almost physical terms: the crossing of a eucultural threshold by the species, followed – perhaps inevitably – by a sustained autocatalytic reaction in which genetic and cultural evolution drove each other forward."

According to Lumsden and Wilson, the human mind as a construct of euculture is represented in *culturgens* – the array of transmissible behavior, mental constructs, and artifacts that denote the basic units of culture. The term *culturgen*, a concept not wholly dissimilar to Dawkins' *meme*, can be defined as generators of culture. The term is derived from *cultura* – Latin for culture and *gen* – Latin for produce. Terms coined by others to correspond to similar concepts include: mnemotype, idea, idene, meme, and sociogene, to name several. Years later, Wilson abdicated the term *culturgen* in favor of Dawkins' *meme*, which had achieved far more regular usage in the academic literature (Wilson, 1998). Regardless of the term presently utilized, Lumsden and Wilson encourage our usage to include within the definition "that the unit of culture – meme – be the same as the node of semantic memory and its correlates in brain activity. The level of the node . . . determines the complexity of the idea, behavior, or artifact that it helps to sustain in the culture at large (Wilson, 1998).

The unique attributes of human culture that distance us from other species are the result of eucultural reification that enables symbolization of memes and their cross-generational transmission. Such reification is observed to occur only infrequently among other primate species. Hypothetically, culturgens may be passed from one generation to the next by means represented along a continuum defined by three possible worlds of behavioral etiology:

(1) *Pure genetic transmission* that would provide a world in which behavioral choices exist but only one will ever be preferred – a kind of innate determinism. Learning is certainly possible in this world as are other behavioral choices, but actual behavior is rigidly channeled by a mind prewired to respond in a stereotypical fashion.

(2) *Pure cultural transmission* that would produce a mind selecting behavior among equally attractive multiple choices – the ultimate blank slate. Individual behavior would be represented by behavioral choices that depend solely on the environment provided by the culture. Those choices would be easily learned, easily transmitted, and fully independent of biological influence.

(3) A mind produced by *gene-culture transmission* is the world we occupy where learning can make a wide range of behavior possible, but where biological predispositions of the brain make specific choices highly probable. This mind is where "genes and culture are held together by an elastic but unbreakable leash. Culture emerges and advances in development by means of innovation, and the introduction of new ideas and artifacts from the outside. However, it is constrained and directed to some extent by the genes. At the same time, the pressure exerted by cultural innovation affects the survival of the genes and ultimately alters the strength and torque of the genetic leash" (Lumsden and Wilson, 1981).

The coevolutionary "elastic leash" image is accurate in its symbolic representation of function, but is a rather unfortunate phrase choice, in that critics use the pejorative connotations of a leash as a means to suggest a kind of biological determinism and control. The emergence of mind as an outcome of gene-culture coevolution can better be represented by a less controlling image, like that of a rubber band in which the flexible rubber band represents genetic predisposition and the environment produces the stimulus forces that stretch the rubber band. The rubber band will always tend toward a particular position of equilibrium but is not as controlling as a leash. Regardless of descriptive image, mind is the operational construct of gene-culture coevolution – what Wilson calls epigenesis. Mind emerges as the sum of all interactions between genes and the culture that achieve expression in the work of neural mechanisms. That existing culture is the "accumulation of a particular history carried in the memories and archives of those who transmit it." We recognize the accumulation of those memes as culture, "a relatively homogeneous group of mental constructions or their products" (Lumsden and Wilson, 1983). At the same time, it is the biological wetware of the mind that encodes the accumulated memes within a generation and prepares to represent (reify) them to the next generation.

Individual members within cultural populations contemplate competing memes in relation to any particular environmental circumstance or combination of circumstances in which they might find themselves. The mind arrives at a result among those competing memes and acts upon it. In reality, the process of coevolution suggests that the "choice" has been constrained by the rubber band of gene-culture transmission and, thereby, limited long ago by the dominant behavioral pattern. Out of the vast number of such "choices" across many categories of thought and behavior, culture grows, is reinforced, or slowly alters its form over time. Each "choice" among a vast array of memes that gives direction to behavioral repertoires combines in a rich interaction to produce the "whole" we observe to be culture. The problem of translating the mind directly to culture is similar to the challenge of determining the whole from the sum of individual parts – understanding the interaction of parts that combine to produce culture is far more complicated. Additionally, "minds do not develop independently of other minds; they are powerfully influenced by the decisions already taken by the rest of the society" (Lumsden and Wilson, 1983).

Additional Considerations

The conventional wisdom of most twentieth century social science is that the contemporary history of our species' behavior has been freed from the "constraints" of biology and is shaped solely by the environment. Wilson challenged that wisdom in a profound fashion. His conceptualization of gene-culture coevolution and its operative mechanism of epigenesis initiated a transformation of the behavioral sciences – a field we know today as evolutionary psychology. Unquestionably, his description of epigenesis as the "interaction between genes and the environment that ultimately result in the distinctive anatomical, physiological, cognitive, and behavioral traits of an organism," defines evolutionary psychology. "But in order to have a real evolutionary theory of mind and culture, one must begin with genes and the mechanisms that the genes actually prescribe. In human beings the genes do not specify social behavior." Genes generate the organic processes called epigenetic rules "that feed on culture to assemble the mind and channel its operation" (Lumsden and Wilson, 1981). Epigenetic rules are apparent as we observe the natural selection of variations in behavior that emerge from the interaction among: (1) the operant shaping provided by the imperatives of environmental consequences, (2) the developing contingencies of human social behavior and culture, and (3) the biological expression in neural information processing modules. The greater biological success of certain kinds of social behavior and culture causes the underlying epigenetic rules and their guiding genes to spread through the population. It is the ongoing coevolutionary interaction of those genes and the environment that drives all phenotypic expression of social behavior and culture. Wilson suggests that this rich gene-culture coevolution achieves its finest expression in the human mind – a continually emerging accumulation of memes that are assimilated via continually emerging epigenetic rules.

The rapidly accumulating research in developmental biology, especially in the newly emerging field of evolutionary developmental biology (evo devo), may hold enormous potential for adding to our understanding of gene-culture coevolution. Evo devo is making rapid progress on explicating the so-called "tool kit" genes and genetic switches, thereby guiding our understanding of the ontogeny of phenotype and epigenetic alterations of gene regulation. The research suggests that expressed phenotypes are not uniquely determined by their genotypes. Rather, the generation of phenotype is dependent on environmental variables that interact with the emerging developmental expression of genotype (Carroll, 2005). Additionally, the extraordinarily extended neotenous condition of the human brain is a variable of notable importance impacting this rich interaction. The work to date has focused on specific observable physical manifestations of mainly fetal development. Future investigations will inevitably explore the emergent physical constituents of the brain that organize the development of behavioral repertoires as an organism moves from its fetal environment into its species-typical environment (Lickliter, 1996). Unquestionably, our better understanding of the evolution of human neural embryology and ontogeny may provide a more complete biological foundation for completing the modern synthesis that leads us to true consilience (Carroll, 2005).

Evo devo is likely to have its most significant influence on our better understanding of human nature by inspiring the advancement of research efforts and conceptual development in evolutionary developmental psychology. As a continuum of interrelated disciplines that emerged in the mid-1980s, the field ranges between cell biology and cultural anthropology, investigating behavioral development by intentionally integrating biological concepts with those from psychology. The clear focus in evolutionary developmental psychology "involves the expression of evolved, epigenetic programs in interaction with an individual's physical and social environment over the course of ontogeny. Central to evolutionary developmental psychology is the idea that there are different adaptive pressures at different times in ontogeny" (Bjorklund and Pellegrini, 2000). Thus, practitioners of evolutionary developmental psychology suggest that studying the discontinuities in the development of an organism's behavioral phenotype will provide insight regarding selection mechanisms that prevail during specific periods of ontogeny – periods that are not necessarily descriptive of behavioral maturation as a seamless continuum from infancy to adulthood.

> *Survival is a problem faced by any organism at any stage of its life span. Because threats to survival can take different forms for organisms of different ages, it is likely that organisms evolve multiple age-specific adaptations for survival during their life span. Therefore, the view that developing organisms may exhibit successive age-specific adaptations and that earlier adaptations may disappear and be replaced by new ones applies well to adaptations for survival.*
>
> *(Maestripieri and Roney, 2006)*

Additionally, a growing body of investigators is pursuing a comparative study focused on nonhuman primates to determine phylogenic relationships among significantly related species. In reality, Darwin (1872) produced a significant body of work describing his observations of human facial expressions of emotions as potentially nonadaptive homologies of animal expressions. Contemporary comparative research is exploring a range of behavioral attributes that include infant's attachment to the caregiver and female interest during the juvenile period (Maestripieri and Roney, 2006).

Consilience

A congenital synthesizer, I held on to the dream of a unifying theory.

(Wilson, 1994)

Perhaps the most compelling and relevant question regarding any serious consideration of sociobiology and its place in the "modern" or "new" synthesis is one that went decades without being overtly discussed – that is, the appropriate resuscitation of group selection as a feature of multilevel selection. Sociobiology has begged the question regarding multilevel selection and group selection from its outset. And, aside from the notable persistence of David Sloan Wilson (1975, 1983, 2002, and 2008) and several others – Elliott Sober and E.O. Wilson (Sober and Wilson, 1998; Wilson and Wilson, 2007; and Wilson, 2008), the early ideological attacks on Wilson's sociobiology from the likes of Gould and Lewontin were replicated by similar ideological attacks on multilevel selection. Recent antipathy has been more precisely focused on group selection, most notably by Richard Dawkins (1976, 1982 and 1997).

Group selection and multilevel selection, by extension, are out of favor – not necessarily on scientific grounds, but on ideological grounds. Sober and Wilson (1998) have reinvigorated the conversation about multilevel and group selection, doggedly persisting in their pursuit of scientific truth and ideological neutrality. They have returned to the original conceptual framework of Darwin and accepted that group selection can explain both the behavior of social insects and virtuous human behavior. Inasmuch as group selection is still credible, so too is multilevel selection. "A growing number of scientists now find it both uncontroversial and highly insightful to think of natural selection as a process that operates on a nested hierarchy of units. Multilevel selection theory is being used to explore an extraordinary range of phenomena, from the origin of life to the nature of human societies" (Wilson and Sober, 1994 and Wilson and Wilson, 2007).

Multilevel selection theory views every level of the biological hierarchy through the same lens of natural selection. "Natural selection occurs when genes differentially survive and reproduce within single individuals, when individuals differentially survive and reproduce within single groups, and when groups differentially survive and reproduce within a global population" (Sober and Wilson, 1998). Indeed, Darwin (1859) originally first offered the concept of multilevel selection by suggesting that "natural selection takes place at more than one level of the biological hierarchy."

Selfish gene proponents criticize multilevel selection and group selection by stating that selfish behavior always prevails. Wilson (2007) resolved that commonly identified dilemma by suggesting that "selfish individuals might outcompete altruists within groups, but altruistic groups outcompete selfish groups." This is the essence of multilevel selection theory, of which group selection is a part. For its part, group selection favors characteristics that enhance the prospects for survival of a group relative to the other group. If group selection acts effectively, the group may evolve into an adaptive unit and be open to study in the same way individuals can be studied. In Darwin's words:

> *There can be no doubt that a tribe including many members who, from possessing in a high degree the spirit of patriotism, fidelity, obedience, courage and sympathy, were always ready to aid one another, and to sacrifice themselves for the common good would be victorious over most other tribes; and this would be natural selection.*
>
> *(Darwin, 1871)*

Wilson and Wilson (2007) go on to urge full re-evaluation of the 1960s rejection of group selection and the basic logic of multilevel selection theory. In fact, "Rethinking the Theoretical Foundations of Sociobiology" (Wilson and Wilson, 2007) argues that such a "back to basics" approach benefits everyone, "from the most advanced theorist to the student learning sociobiology for the first time." In closing, they submit that:

> *group selection is an important force in human evolution in part because cultural processes have a way of creating phenotypic variation among groups ... If a new behavior arises by cultural mutation, it can quickly become the most common behavior within the group and provide the decisive edge in between-group competition ... The importance of group selection in human evolution enables our groupish nature to be explained at face value ... Thus, multilevel selection, not group selection alone provides a comprehensive framework for understanding human evolution along with other major transitions.*
>
> *(Wilson and Wilson, 2007)*

Certainly, "multilevel selection theory, including group selection, provides an elegant theoretical foundation for sociobiology in the future (Wilson and Wilson, 2007)." This most recent work by Wilson is critical to the continued vitality of sociobiology and clearly represents Wilson's expansive and synthetic intellectual style. It also represents an extremely useful conceptual and empirical pursuit at this point in his distinguished career – a pursuit that rejects the most naïve form of group selection that accepts that behavior evolves for the good of the group. Rather, "traits with public benefits and private costs do evolve by natural selection ... the balance between levels of selection needs to be evaluated on a case-by-case basis" (Wilson and Wilson, 2008).

Contemporary sociobiology has much to gain by Wilson's twenty-first century contributions. Many scholars in the field had long anticipated additional seminal contributions from him after a long absence of his sociobiology voice. Following the publication of *Promethean Fire,* Wilson retreated from active publication regarding sociobiology in favor of more aggressive pursuit of his lifelong interest in biodiversity and his newly emerging concept of biophilia (1984). In fact, it is far more likely that recent generations of biologists are acquainted with Wilson for his work in biodiversity than for his work in sociobiology. After nearly 15 years he exercised his sociobiology voice again; *Consilience: The Unity of Knowledge* (1998) was the result. By "unity of knowledge" he intended to challenge the entire intellectual community, including the humanities and fine arts, to become a part of a comprehensive intellectual synthesis that pushed beyond the "new synthesis" of his sociobiology. He proposed that this new "consilience" would achieve a complete knowledge of human nature as it aligned the collective wisdom from each discipline. Further, he proposed that the thread to stitch the fabric into one piece should be a deep understanding of gene-culture coevolution.

Wilson's consilient outcome would be a community of scholars steeped in the wisdom of their own discipline and well-informed about the biological basis of human nature. Such interdisciplinary and synthetic insight would enable the larger academic community to assume a more prominent role in educating humanity to achieve a sustainable future. Until this community of scholars is achieved, "we will continue to drown in information, while starving for wisdom. The world henceforth will be run by synthesizers" (Wilson, 1998). Consilience is a fascinating prospect and an admirable goal – the book is commendable for offering a lucid rationale to that end. Scholars have variously agreed and disagreed whether Wilson was successful in achieving his ambition for the book (Henriques, 2008).

A more pragmatic result accomplished by Wilson in *Consilience* was its contemporary recapitulation of sociobiology after a decade and a half of near silence. That passage of time enabled him to achieve a simple elegance in summarizing the essential principles of gene-culture coevolution for a new generation of readers. The two-time Pulitzer Prize winner is always best represented in his own words. Thus, the following passages are offered from *Consilience* to provide an appropriate summary and closing to this chapter.

> *Culture is created by the communal mind, and each mind in turn is the product of the genetically structured human brain. Genes and culture are therefore inseverably linked. But the linkage is flexible, to a degree still mostly unmeasured. The linkage is also tortuous: Genes prescribe epigenetic rules, which are the neural pathways and regularities in cognitive development by which the individual mind assembles itself. The mind grows from birth to death by absorbing parts of the existing culture available to it, with selections guided through epigenetic rules inherited by the individual brain.*
>
> *As part of gene-culture coevolution, culture is reconstructed each generation collectively in the minds of individuals. When oral tradition is supplemented by writing and art, culture can grow indefinitely large and it can even skip generations. But the fundamental biasing influence of the epigenetic rules, being genetic and ineradicable, stays constant.*
>
> *Some individuals inherit epigenetic rules enabling them to survive and reproduce better in the surrounding environment and culture than individuals who lack those rules, or at least possess them in weaker valence. By this means, over many generations, the more successful epigenetic rules have spread through the population along with the genes that prescribe the rules. As a consequence the human species has evolved genetically by natural selection in behavior, just as it has in the anatomy and physiology of the brain.*
>
> *The nature of the genetic leash and the role of culture can now be better understood as follows. Certain cultural norms also survive and reproduce better than competing norms, causing culture to evolve in a track parallel to and usually much faster than genetic*

> *evolution. The quicker the pace of cultural evolution, the looser the connection between genes and culture, although the connection is never completely broken. Culture allows a rapid adjustment to changes in the environment through finely tuned adaptations invented and transmitted without correspondingly precise genetic prescription. In this respect human beings differ fundamentally from all other animal species. (pp. 127–8)*

These steps can be summed up very briefly as follows:

> *Genes prescribe epigenetic rules, which are the regularities of sensory perception and mental development that animate and channel the acquisition of culture.*
>
> *Culture helps to determine which of the prescribing genes survive and multiply from one generation to the next.*
>
> *Successful new genes alter the epigenetic rules of populations.*
>
> *The altered epigenetic rules change the direction and effectiveness of the channels of cultural acquisition. (p. 157)*

In closing:

> *The brain constantly searches for meaning, for connections between objects and qualities that cross-cut the senses and provide information about external existence. We penetrate that world through the constraining portals of epigenetic rules ... In order to grasp the human condition, both the genes and culture must be understood, not separately in the traditional manner of science and the humanities, but together, in recognition of the realities of human evolution." (p. 163)*

Resolving the origins of culture may go a long way to resolving the problem of mind. The prospect may exist that our best operational descriptors of mind exist at the intersection of paradigms advanced by Wilson and Skinner. In that context, mind is an emergent expression of the epigenetic process produced by the accumulation of memes that animate epigenesis *and* the feedback mechanism of environmental contingencies (operant shaping) of the behavioral choices considered by the individual. Our vocabulary is inadequate to provide accurate expression to this concept. We are reduced to utilize *mind*, with all its phenomenological and definitional baggage, and *choices*, with all its connotations of conscious action, freedom, and intent, in order to avoid the claims of reductionism and determinism.

With little doubt, mind is a word that represents a biological mechanism that emerges with the interaction of Skinner's three kinds of selection by consequences and their combined relatively proximate and relatively ultimate causal mechanisms. Darwin suggested the possibility that "natural selection has been the most important, but not the exclusive, means of modification." Those closing words from the introduction to *The Origin of Species* anticipated the prospect that so complex a phenomenon – the mind, might require the explanatory power of Skinner's operant selection by consequence, Wilson's gene-culture coevolution partly driven by that operant selection, and an inspired renewal of multilevel selection theory and group selection within the added context of Skinner's three kinds of selection by consequences.

Chapter 3
A Consilient View of B.F. Skinner and E.O. Wilson: The Operant Foundation of Sociobiology

Great is the power of steady misrepresentation; but the history of science shows fortunately this power does not long endure.
(Charles Darwin, 1859)

B.F. Skinner and E.O. Wilson are two of our most influential thinkers regarding human nature since Charles Darwin. Each provided remarkable scientific insight regarding the essential nature of behavioral variation and selection in all living organisms, including humans. Skinner's operant conditioning has motivated generations of scientists to explore the full effects that scientifically manipulated consequences have on the acquisition and maintenance of behavior. Wilson's sociobiology describes the biological basis of social behavior and culture; it has encouraged vigorous scientific discourse regarding its explanatory power and has motivated the development of the field we know today as evolutionary psychology. Their careers as faculty colleagues at Harvard spanned decades. They maintained cordial relations during that time and held mutual high regard for each other's work.

True to form, the intellectual community responded in a bipolar fashion to their works. Skinner's radical behaviorism elicited powerful responses for its straightforward empirical effectiveness demonstrating the operant basis for behavior. He also provoked enormous angst among many who viewed his operant conditioning techniques as reductionistic, at best, and crudely manipulative, at worst. Wilson's sociobiology prompted widespread acceptance within the biological sciences for its power describing the evolution of social behavior and culture. At the same time, it also achieved widespread scorn among many social scientists and some in the scientific community as reductionistic, deterministic, and a scientific justification for everything bad in human society.

Nonetheless, shifting paradigms that have an effect on the prevailing intellectual, political, philosophical, religious, and cultural ideologies always threaten the delicate balance of human understanding and challenge theorists to reconsider intellectual convictions. The dissonance caused by Skinner's and Wilson's new paradigms provoked generations of thinkers to reassess their former conceptions of behavior and, for many, fundamentally affected the way they understood the human place in the great chain of being.

P. Naour, *E.O. Wilson and B.F. Skinner*, Developments in Primatology: Progress and Prospects, DOI 10.1007/978-0-387-89462-1_3,

Much has been written and even more said regarding the important work of Skinner and Wilson. However, no serious intellectual attempt has been made to locate and explore significant complementary relationships in their work and suggest the potential compatibility of their thinking. An alignment of the essential character of their thought into an all-inclusive paradigm is now warranted.

Even though Skinner enjoyed enormous attention in the behavioral sciences community, his work remained largely unknown to the emerging field of evolutionary psychology and the larger community of biological scientists. Skinner's dogmatic "black box" approach and self-described radical behaviorism made it highly unlikely that any biologist would be drawn to, much less seriously consider, the potential biological implications of his theoretical framework. He maintained that hard-nosed radical disposition until very late in his career.

Less than three years before his death, Skinner joined his colleague, E.O. Wilson in a conversation. It is quickly apparent as they begin their discussion that one of Skinner's obvious intentions was to place himself in a much larger scientific context and to have his work considered for its evolutionary and biological implications. Skinner moved the early conversation through biological concepts of selection, organization, and complexity and made reference to the important work of Ernst Mayr. He quickly went on to comment on culture and Wilson's sociobiology, suggesting, "Sociobiology seems to leave me out. I'm in the middle." Wilson replied by recalling C.H. Waddington's *New York Review of Books* (1975) essay on Wilson's newly published *Sociobiology*, in which Waddington suggested, "he (Wilson) left out the mind." Skinner quickly responded, "Yes, I think so. I think mind, person, or individual have to come into it." On its own merit, this one comment is a remarkable and wholly unanticipated acknowledgement of "mind" by the preeminent radical behaviorist; it sets the stage for a fascinating intellectual departure by Skinner as he invited Wilson to peer into the "black box" with him.

When Skinner suggested to Wilson that "Sociobiology leaves me out. I'm in the middle," he implied that the field generally failed to recognize the power of operant conditioning in the larger context of evolutionary theory as it relates to the phenotype of behavior. Certainly, sociobiology recognizes Darwin's natural selection as a critical relatively ultimate cause mechanism. However, sociobiology does not recognize what Skinner describes as the two other kinds of selection by consequences – real-time consequences responsible for selecting individual behavior and special consequences maintained by the social environment (Skinner, 1981). Skinner viewed himself in the middle between natural selection and sociobiology. He saw operant mechanisms as the bridge extending from natural selection to the coevolutionary mechanisms described by Wilson that lead to culture. The power of operant conditioning as a biological agent driving behavior was apparent to Skinner. The absolute responsiveness of all organisms to be operantly conditioned is a product of neural evolution. Additionally, the operant contingencies selecting individual behavior and those maintained by the social environment shape the course of naturally selected, yet still malleable, neural circuits during individual ontogeny. His challenge was to lead Wilson to understand the relationship of this reality to sociobiology and the evolution of social behavior and culture.

The perennial difficulty in this scientific arena has been to clarify the distinctions between the more apparent physical manifestations of evolution and the more subtle behavioral manifestations of evolution – often discussed among evolutionary biologists as the phenotypic morphology versus phenotypic behavior paradox. The more mechanistic selection exerted on the evolution of physical attributes occurs as environmental pressures have bearing on the capacity of organisms to advance genes to subsequent generations. These physical characteristics are most typically overtly observable, have apparent relationship to the environment, and are obvious in the fossil record. The more subtle selection of behavioral repertoires has its physical manifestations only in the "black box." Skinner's comments implied that neural capacity for absolute responsiveness to operant conditioning was selected just as physical attributes were selected; in this case, the "black box" is the physical manifestation. Regardless of phylogenic sophistication, all organisms capable of emitting behavior are operantly conditioned and shaped by consequences. Thus, Skinner provided biology an intriguing model to explain how organisms are behaviorally adaptive to rapidly changing environmental conditions. The link between the slow-paced natural selection of physical attributes (including brain) and intraorganism behavioral adaptation was accomplished by natural selection of neural tissue that compelled organisms to operant conditioning.

Consequences play an essential role in both the selection of physical characteristics and behavioral repertoires. For Skinner, consequences result in increasing the probability that today's behavior will continue or cease; for Wilson, consequences drive selection and result in an organism's capacity to survive long enough to pass genes to the next generation. Skinner's focus was immediate and within the organism – behavior acquisition, rates of behavior, maintenance of behavior, resistance to extinction, and behavioral shaping. Wilson maintained an evolutionary perspective which assumed that consequences have a long-term multigenerational influence – dispositions toward shared social behavior repertoires within a species that advance the survival of the group and combine to produce culture. Skinner's long-term oversight was neglecting to claim the biological nature of the "black box" as the ultimate source of free operant behavior. His overture to Wilson made apparent his desire to remedy his negligence.

Skinner attempted to prompt Wilson to consider the possibility that organisms possessing the capacity to be operantly conditioned had survival advantage during the very earliest natural history of life on earth. Organisms that were predisposed to appropriate responsiveness to the subtleties of environmental contingencies were more likely to pass that tendency to offspring. He did not suggest that a particular stimulus caused a determined response. Skinner suggested a very malleable, yet powerful, biological capacity to be more efficiently shaped in real-time by environmental consequences. Specific behavior and behavioral repertoires that resulted from ever-present operant shaping become the common behavioral traits within that organism and, likely, across the species over time. But, they are not mechanistic stimulus/response behavior; they are emitted behavior shaped and canalized by the contingencies of the environment and increasingly organized as behavioral predispositions in the developing neural circuitry of the "black box." Indeed, the "black box" is neither tabula rasa nor equipotential. The brain is a magnificent tapestry of neural modules that drive behavioral expression – modules that have been operantly shaped by the contingencies of environmental pressures over individual ontogeny and evolutionary history.

Might it then be possible to expect that behavioral patterns within a species act upon malleable neural circuits early in development to become more predisposed to and increasingly efficient at emitting that behavior? Yes. Might it also be possible that early in a species' evolutionary history, the circuits responsible for categories of neural processing were naturally selected? Yes. In other words, neural plasticity is shaped during ontogeny by relatively proximate cause selection imperatives of operant conditioning. Thus, for a much oversimplified example, neural tissue in rats with the capacity to process sequential information was operantly shaped by environmental contingencies to process visual-spatial information in a sequential fashion that, among other behavior, enabled rats to efficiently negotiate return to the safety of a nest. This same sequential neural processor in a more sophisticated human brain was operantly shaped by the selecting consequences of the environment to process the sequential auditory elements of language. Linguistic advancement in early hominids was a result of more efficient sequential neural capacity. It enabled them to more successfully process auditory units in proper sequence to create meaning. In both rats and humans, those individuals with more efficient and productive sequential neural capacity were at a selection advantage and were more likely to pass that capacity to offspring. The continuum of relatively proximate and relatively ultimate causality represented in this example of behavioral expression of neural evolution is precisely the kind of naturalistic study that illustrates the importance of explanatory relativity described by Sober and Wilson (1998) and more fully developed by Amundson (2005).

The time is right to appreciate the magnitude of Skinner's intellectual contributions that place him among the elite thinkers of modern science – contributions that go beyond operant conditioning. He was wise to draw Wilson into a most uncommon conversation in which both acknowledge the concept of mind as a biological entity. In order to better follow their conversation (Chapter 4) it will be useful to review the essential nature of Skinner's work in relation to some basic concepts of Wilson's sociobiology.

Radical Behaviorism

Skinner's operant conditioning paradigm empirically describes how all emitted behavior of organisms is shaped by the positive and negative consequences to those operations on the environment. Immediate benefits (positive reinforcers) like food and safety increase the likelihood a specific behavior will recur, while behavior resulting in negative (punishment) or no consequences will likely not recur. Organisms "learn" to emit behavior that have potential benefit and avoid that behavior having potential loss. Obviously, the natural environment produces real world consequences to behavior that are most significantly beneficial – food, safety, and access to mates; she also provides the harshest consequences – separation from the group, hunger, thirst, no access to mates, and possible death. Skinner used the term primary reinforcer to label those that are necessary to sustain life, while the term secondary reinforcer is used to identify those that require learning.

The operant paradigm was a bold departure from the mechanistic and involuntary causal relationships identified among unconditioned stimuli, unconditioned responses, conditioned stimuli, and conditioned responses in Pavlov's classical conditioning experiments of the late nineteenth and early twentieth centuries. Skinner also advanced the important transitional work of Edward Thorndike, who identified two primary behavioral principles: (1) the law of effect and (2) the law of multiple responses. The law of effect states that learning is a function of the consequences of behavior; pleasant consequences increase a behavior and unpleasant consequences diminish or stop behavior. The law of multiple responses says that organisms will engage in a variety of behavior until one produces a satisfying effect.

Skinner ingeniously utilized the foundational work of his behaviorist predecessors – most notably Thorndike – and described the conditioning (effect) of "free operant" behavior (multiple responses) produced by organisms in relation to their environment. He then spent the bulk of his scientific career investigating the variables that affect operant conditioning. He demonstrated that organisms engage a wide variety of free operant behavior that have no identifiable causal relationship to stimuli produced in the environment. Instead, an organism emits free operant behavior until a positive consequence reinforces the behavior and thereby increases the likelihood the behavior will be emitted again.

Radical behaviorism was Skinner's lifelong intellectual obsession. Perhaps most radical in the view of critics was his confident assertions that operant conditioning explained the basis for all "voluntary" behavior and that "free will" was just an illusion. Find fault with the boldness of those assertions? Yes. Find holes in his empirical practice? Never. His comprehensive research spanning decades, consistently demonstrated the efficacy of each component of the paradigm. Carefully identify a target behavior, positively reinforce the behavior to assure its acquisition, and then apply the appropriate schedule of reinforcement to assure its resistance to extinction. He described these schedules as they relate to the natural contingencies of our world and to the structured contingencies of our managed world as authorities apply reinforcement schedules. Simply summarized, the behavioral repertoire of the simplest to the most complex organisms is the accumulation of a lifetime of operant shaping and reinforcement history.

Most powerful among his contributions regarding reinforcement was the concept of negative reinforcement – an organism emits an appropriate behavior in order to avoid the prospect that something bad might occur. The subtlety in negative reinforcement requiring careful explanation is that it sustains appropriate behavior and is not to be confused with punishment that diminishes behavior. Skinner argued that the majority of human behavior is sustained by negative reinforcement, otherwise referred to as "guilt" by many of us. Likely, his careful exploration of the powerful effects of negative reinforcement led him to focus significant attention on ethical considerations later in his career. *Beyond Freedom and Dignity* (1973) addressed many of his most serious ethical concerns, in addition to a larger philosophical consideration of free will.

Skinner was profoundly cognizant of the inherent power in careful application of operant conditioning techniques, both for good and evil. He gave significant attention in his writing to guide ethical considerations posed by his behavioral technology. He argued for positive application of his operant techniques and warned of the misapplication of both positive and negative reinforcement, and of the moral considerations necessary in the applications of punishment. *Walden Two* was Skinner's utopian novel in which he attempted to show how operant principles could be applied to organize an entire society, while *Beyond Freedom and Dignity* was his more useful philosophical contribution in which he challenged us to consider the "fiction" of free will and the faulty logic of basing entire social structures on its assumed existence. Throughout his life, he unequivocally maintained that operant conditioning is effective and that our more important task was to understand it and manage it to positive outcomes, rather than to blindly attempt to frame ideological arguments diminishing its efficacy. Perhaps a bit hyperbolic in his claims, but nonetheless certain to cause his readers pause as they consider applying his paradigm to better understand behavior.

According to Skinner, positive consequences (or the avoidance of negative consequences) that condition us to intraspecies social structures (culture) are among the most significant reinforcers shaping large classes of behavior. Species have survival advantage when they act within and on behalf of their common (perhaps "selfish") genes. Skinner describes these consequences as group contingencies, meaning that all members of a group achieve a positive outcome when individual behavior conforms to expectations of the group. In reality, Skinner described the array of within group behavior most likely to be shaped as a result of group contingencies as those that benefit the whole group. In a real sense, Skinner's operant paradigm describes the natural selection of individual and group behavioral repertoires. Environmental consequences operantly shaped by differential reinforcement provide the individual and the group with survival advantage. As well, the group provides models of that behavior that can be imitated during initial stages of behavior acquisition.

Operant shaping of imitated behavior through differential reinforcement, an elusive concept to many, is an essential component to the natural selection of increasingly sophisticated human behavior. Skinner often used the example of human language acquisition to demonstrate naturally occurring operant shaping of imitated behavior. Consider the young child just beginning to babble and the nature of parental response to that child. Initially, the parents' reinforcement of the babbling behavior is offered with little or no expectation of behavioral competency. Along with enormous attention and praise (a primary and a secondary reinforcer), the parent is typically mouthing and speaking back to the child (cuing the child to imitate). This pattern of parental response both conditions the initial "primitive" language behavior of the child, while modeling and cuing the child to improve their articulation patterns. This increasing behavioral expectation most typically is accompanied by reinforcement contingent upon the child's incremental improvement. Skinner labeled this deliberate increase of behavioral expectations and contingent reinforcement as operant shaping by utilizing differential reinforcement.

Skinner gave his professional life to advance our understanding of the most productive and efficient application of applied behavioral analysis and operant conditioning. The concepts of primary and secondary reinforcement; schedules of reinforcement, shaping, and the application of differential reinforcement; resistance to extinction; and imitation are among his most important contributions. His radical behaviorism had an enormous impact on twentieth century American psychology and on the many areas of applied psychology, especially American education. However, he remained largely neglected by the biological community because he maintained the irrelevance of the brain as a meaningful basis for understanding the intricacies of behavior. Similarly, he never offered a comprehensive assessment of the biological nature of survival advantage as an outcome of behavioral selection, nor did he push beyond the important concept of three kinds of selection by consequences that he introduced in 1981.

Skinner's final scholarly contribution, completed the day before his death, more completely conceptualized much of what he was attempting to frame in that 1981 article and in the conversation to which he invited Wilson in 1987. A careful reading of that final work (*Can Psychology Be a Science of Mind?*), makes apparent that Skinner had given considerable thought to the larger biological issues of behavioral selection following his conversation with Wilson. It is also quickly apparent that Skinner had utilized that conversation with Wilson to think out loud, while attempting to guide Wilson to consider the larger biological implications of the operant paradigm and perhaps encourage him to weave some of his own thinking regarding selection by consequences into his subsequent work. That prospect never came to fruition nor has Skinner's final conceptual contributions regarding variation and selection of behavior achieved much attention since his death. Skinner is now due that larger audience.

Perhaps Skinner was insistent about the "black box" to a fault, disregarding any reference to covert biological processes and thereby alienating a considerable body of scholars in the life sciences. Had he more effectively framed his operant conditioning paradigm within the larger context of biological systems and the variation and selection of behavioral dispositions earlier and more regularly in his career, a far larger scientific community likely would have included his paradigm in their thinking. Had Skinner described operant conditioning as a natural biological imperative innately functioning within all living organisms, ethologists and biologists might have easily seen the mechanisms as the common means by which all "voluntary" behavior is shaped and all organisms are innately responsive to the operant contingencies in the natural world. In other words, the capacity for an organism to be responsive to operant conditioning and thereby emit behavior appropriate to an environment has been naturally selected. The simple fact that the environmental consequences systematically described in Skinner's operant model universally shape all "voluntary" behavior in organisms is a straightforward endorsement of its natural selection and survival value.

Skinner would have benefited by including the thinking of his contemporaries, D.O. Hebb and Paul MacLean, in his early formulations regarding operant conditioning. Hebb's, *The Organization of Behavior: A Neuropsychological Theory* (1949), is considered a classic to this day because he combined a studied behavioral approach with biological conceptualizations regarding neural "cell assemblies" and wide-ranging considerations regarding the brain basis for learning and intelligence. MacLean's triune brain model, first conceptualized in 1952 and fully explicated in 1989, provides an evolutionary model for the phylogenic advancement of behavioral capacities organized within three distinct brain units. Both contemporaries might have motivated more acknowledgement of the emerging work that began to open the "black box."

MacLean's work, in particular, is suggestive of biological mechanisms that had been so effectively described in behavioral psychology. MacLean's most primitive reptilian brain is not capable of learning, but is innately hardwired to subserve basic survival functions of an organism that are released by environmental stimuli. The more recently evolved paleomammalian brain provides involuntary emotional content to an organism's behavioral repertoire and can be classically conditioned in the traditional Pavlovian sense. The most recent evolutionary development is the neomammalian brain which provides "intelligent" cognitive modules enabling an organism behavioral adaptability. An intriguing challenge for Skinner would have been to identify how MacLean's three levels of brain organization and their related functions are the biological expression of his three kinds of selection by consequences (1981). All three levels of MacLean's model are obviously an expression of natural selection, with the paleomammalian brain representing basic instinct. Individual operant contingencies likely have the greatest impact on shaping the paleomammalian brain, while the neomammalian brain are most likely shaped by the consequences of selection maintained by the social environment.

Aligning Skinner and Wilson

... naturam animi atque animai corpoream ...
(the mind and spirit have a material nature – Lucretius, 50 BCE)

The operant conditioning and behavioral shaping Skinner so precisely described as a "second and third kinds of selection by consequences" provide the bridge from Darwin's natural selection to Wilson's sociobiology (Skinner, 1981). Skinner produced little additional writing beyond that seminal 1981 article that provided any significant insights in that regard nor did he often extend the range of his intellectual pursuits beyond radical behaviorism, its applications, and its ethical implications. In comparison, the bulk of Darwin's work focused on the evolutionary outcomes suggested by the more observable natural selection of physical attributes that increased the survival prospects of an organism. He gave far less attention to the evolution of behavioral attributes in his work, aside from discussions of habits and instincts. And, regardless of the essential scholarship that led to Wilson's *Sociobiology: The*

New Synthesis (1975), it was not until its publication that evolution of culture and social behavior received full consideration. The possible relationship among Darwin's natural selection, Skinner's behavioral selection, and Wilson's coevolutionary approach has never been meaningfully established or explicitly explored.

Wilson's paradigm is based upon decades of research on insect societies (1971) and the biological basis for their evolution as group behavior advances its survival prospects. In his final chapter of *Sociobiology* – "Man: From Sociobiology to Sociology," he asserts that all species, including humans, are guided by the same biological imperatives of natural selection codified in their genes for common social behavior. Unlike Skinner, who gave meticulous attention to the environmental basis for individual behavior, Wilson developed his model with a focus on group behavior most frequently categorized as social behavior or culture. He posits that those "cultural norms" are the product of generations of natural selection among repertoires of behavioral predispositions inherently organized within the tissue of the brain as codified by the human genome. He never suggests that the genome predisposes humans to stereotypical stimulus/response-like patterns of behavior. Rather, he goes to great lengths to describe the enormous range of so-called "normalcy" among humans as counter posed by the neurologically simpler insect societies. He asserts that humans continue to select from among ranges of behavioral dispositions as our species continues to emerge in the face of ever-changing environmental contingencies. Humans have remarkable plasticity in behavioral response to environmental conditions. "Some increase in it (plasticity) is to be expected ... What is truly surprising, however, is the extreme to which it has been carried" (Wilson, 1975).

Wilson's sociobiology unequivocally argues for a biological imperative in the emergence of human culture and social behavior. Behavioral repertoires such as courtship and marriage, family, spirituality/religion, altruism, war, and consent to govern are all examples of behavioral predispositions shaped by environmental contingencies during our evolution. Importantly, he never asserts that behavioral predispositions that served at one time in our evolutionary history are by necessity good in our contemporary era. Tribal conflict and intergroup discrimination might have served an essential evolutionary function in primitive times when the consequence of these behavioral dispositions was survival of the gene pool to the next generation. The enormous magnitude of the wars and the racial and ethnic discrimination of our contemporary era bear no relationship to the selection pressures of the ancient African savannah. However, those behavioral predispositions have their biological foundation in that ancient world.

Understanding the primitive impulse toward dispositions of destructive social behavior can go a long way toward better understanding how we might more effectively invoke behavioral plasticity for a more positive result. "Human societies have effloresced to levels of extreme complexity because their members have the intelligence and flexibility to play roles of virtually any degree of specification, and to switch them as the occasion demands" (Wilson, 1975). Yet, the behavioral choices are inevitably and predictably controlled by the variable effectiveness among the consequences to each choice. Sadly, it is too often the case that the most powerful consequences are controlling maladaptive social behavior. This result most frequently occurs among homogenous groups as they relate to other homogenous groups – be they ethnic groups, racial groups, religious groups, or the range of socioeconomic groups.

Evolutionary imperatives are shortsighted and absent values in the face of selection pressure. Wilson submits that our better understanding of humans' genetic heritage in behavioral dispositions might guide our understanding of, but not justification for, human behavior that appall us today. Culture and social behavior are neither good nor evil; however, the outcomes of behavior defined as cultural may be perceived as good or evil in particular settings and in particular times. Culture and social behavior are biological imperatives guided by natural selection. Genes that code the relatively ultimate cause leading to the flexible behavioral repertoires of culture and social behavior have been selected and passed on, while other genes have been lost from the pool. What we witness in our day-to-day living directly corresponds to that ultimate cause – the capacity to be operantly conditioned. More typically, relatively proximate cause explanations such as ethnic or national identity, religion, or race have been offered to describe the basis of cultural and social behavior.

Wilson's overarching framework of sociobiology advances the key concept of gene-culture coevolution as the basis for the evolution of culture and social behavior. *Genes, Mind and Culture* (1981), coauthored with Charles Lumsden, provided an in-depth description of this process and provided mathematical models. Wilson made that same concept more accessible to the nontechnical reader in *Promethean Fire: Reflections on the Origin of Mind*, published in 1983. Both books introduce and develop the concept of culturgens (memes) as the basic units of a culture – units that can be passed among individuals, replicated, or modified that include artifacts, behavior, and mental constructs. The conceptual relationship of Wilson and Skinner can be cultivated within the construct of Wilson's gene-culture coevolution (relatively proximate cause) and Skinner's demonstration of the efficacy of operant conditioning (relatively ultimate cause).

Skinner's radical behaviorism and operant conditioning paradigm were absent in Wilson's conceptual framework. And, Skinner had rarely made claims to biological application, much less evolutionary relevance. Skinner's disinclination to engage in conversations outside the bounds of radical behaviorists to include ethologists and evolutionary biologists had significant implications to the more far-reaching legacy he had earned – far beyond the scale of his behaviorist legacy. He recognized the magnitude of this intellectual parochialism very late in his career. Less than three years before his death, Skinner sought a conversation with Wilson to explore the biological implications of operant conditioning and the possible alignment of their thinking. The transcript of their conversation makes apparent Skinner's desire to help Wilson understand that organisms' innate susceptibility to be operantly shaped by natural environmental contingencies was an early and necessary biological adaptation, and the relatively ultimate cause of culture and social behavior.

Despite these richly suggestive points of interface, neither man ever initiated a regular routine of intellectual discourse with the other in an effort to consider mutual paradigmatic linkages during their many years together as faculty colleagues. Suddenly, the opportunity that had escaped them across the decades of their illustrious careers manifested itself in November of 1987. Their dialogue sets the stage for aligning the two paradigms within a more comprehensive system that includes insight regarding the relationship of operant shaping to the biology of cultural development. To be sure, there is a meaningful paradigmatic linkage between the two. Wilson should apply the operant conditioning paradigm to show how individual and group selection by consequences shape the behavioral repertoires of culture. In order to make that application he must first acknowledge that susceptibility to operant conditioning is a primitive biological imperative (relatively ultimate cause) for all organisms that can emit behavior, while the evolution of neural circuits that provide for the cognitive modules responsible for social/cultural behavior are more recent biological phenomena (relatively proximate cause) that are shaped by gene-culture coevolution.

Operant conditioning is a naturally occurring biological process by which behavior is shaped through environmental contingencies. Organisms have evolved this innate susceptibility to be operantly conditioned as an outcome of natural selection. Thus, the neural organization of behavior is wired with the necessary structures making operant conditioning a biological process in all organisms – none are exempt. Remarkably, this biological capacity to be operantly shaped provides organisms an efficient mechanism for rapid behavioral adaptation to pressures within their environments. In itself, operant conditioning is a natural selection of sorts – a behavioral selection that occurs over much shorter time frames than biological selection. Organisms operate on their environments (Skinner's free operant behavior) by emitting behavior more likely to result in a positive outcome. Should a positive outcome result (Skinner's positive reinforcement), the behavior is more likely to occur again – in other words, it has been "naturally selected." Emitted behaviors that are maladaptive will result in negative consequences and a reduced likelihood the behavior will recur – for Skinner, extinction (a very real prospect in the natural world).

By extension, Skinner suggested that behavioral repertoires viewed as social and/or cultural are products of reinforcement history and operant shaping within a group or species. The relatively ultimate cause of all "voluntary" behavior in Skinner's view is biological susceptibility to operant conditioning. The biology of neural circuits naturally selected to serve specialized behavioral capacities (culturgens) are relatively proximate causes of behavior, which provide for "the evolution of the psychological mechanisms that comprise the human capacity for entering into culture" (Plotkin, 2000). Skinner would likely argue that the brain basis for the sophisticated auditory and speech processes of human language came as a result of environmental contingencies taking advantage of malleable neural tissue. That neural tissue was relatively proximate causation; the biological imperative providing for susceptibility to operant shaping is relatively ultimate causation.

Were he alive today, Skinner would likely accept that contemporary researchers seeking to identify neural circuitry for spirituality, for example, might very well discover it. He would also quickly add that the behavioral predispositions to emit the array of behavior that result in behavioral patterns we describe in the present era as spirituality were operantly conditioned and took advantage of malleable neural tissue – operant conditioning as relatively ultimate cause, religious behavior (culturgens) related to neural circuitry as relatively proximate cause. The behavioral repertoires of culture are biologically served in neural circuits that are proximate causes of that behavior; the biological basis for susceptibility to operant conditioning is the ultimate cause. Had there been an array of behavior other than religiosity having similar or stronger survival value, the neural tissue might as easily been given over to that.

A crucial and controversial feature in understanding the biological basis for culture must be responsive to the claim that it is deterministic in its reductionism. Much of the angst over determinism would have likely been avoided if Skinner had been less radical about his insistence on ignoring the "black box." His detractors' claims of causal determinism in cultural behavior could have been minimized had he allowed for considering the plasticity within the neural substrates required for organisms to emit those behaviors. Instead, criticisms that emanated from that naïve stimulus/response perspective proliferate to this day.

More sophisticated "intelligent" organisms are not constrained to stereotypical responses – biological circuitry does not suggest a determinism that requires prejudice or injustice for instance. Intelligence is the capacity of any organism to emit behavior organized within sophisticated, yet malleable, neural circuits that allow for behavioral flexibility (Plotkin, 2000). Increasing cortical size and increasing periods of juvenile brain maturation (neoteny) allow for neural plasticity that provides for a range of behavioral expressions within the biological predispositions made possible by a "domain specific module" (Plotkin, 2000). Thus, a contemporary expression of religious behavior (culturgen) or verbal behavior (culturgen) may be quite different from that of generations long past. Similarly, contemporary expressions of cultural behavior in diverse environments may well be distinctive. Plasticity, neural flexibility, and extended ontogeny are the hallmarks of big brains. Biologically selected behavioral predispositions provide the trajectory; operant conditioning shapes uniqueness both within individuals and in cultural expression. Research on developmental plasticity and evolution holds interesting potential for explicating the molecular mechanisms that provide for this biologically and environmentally driven emergence of culture as a behavioral phenotype (see Chapter 5).

The Dilemma of Free Will

The real problem in our best understanding of the basis for the evolution of social behavior and culture is not in the ideological charge of determinism, but in the nearly universal and strongly held belief in "free will" as foremost among the cultural expressions of behavior considered essential to human social behavior and culture. Indeed, could the commonly held concept of culture work for most humans if "free will" was nonexistent? The chimera of "free will" has been an enormous impediment to an empirically sound exploration of the biological basis for social behavior and culture. Sadly, the illusion of humankind's culture and "free will" somehow distinguishing it from the great chain of being has betrayed the empirical integrity of a number of otherwise brilliant contributors to this field. Skinner's notion of free will continues the philosophical tradition of Spinoza who, in the seventeenth century suggested that free will is nothing more than a perception like that of stone achieving consciousness upon being tossed in the air and perceiving it was flying of its own volition (*Ethics*, Part I, 1677).

Skinner argued in *Beyond Freedom and Dignity* that humans' sense of worth is to earn dignity by "freely" choosing to engage in culturally acceptable behavior. He provides ample evidence for the widespread acceptance of free will – most notably, legal systems are based on the premise that humans "freely" choose to engage in civil or criminal behavior. The illusion of free will must be maintained in order for humankind to view itself as relevant in the great chain of being. However, the actual expression of free will is strictly constrained to extremely narrow windows and is shaped by the everyday consequences that are inherently a part of our environments – yet many persist in assuming the far-reaching nature of our free will. Living organisms do emit behavior rather than doing nothing; that very act of doing something is a primary reinforcer. Beyond this basic expression of "voluntary" behavior, Skinner argues that contingencies of reinforcement are in control.

Ubiquitous environmental consequences shape social behavior – consequences provided by parents, family, neighbors, and community. However, that same environment is crucial in setting the stage for that behavior to be emitted. All behaviorists, including Skinner, have recognized the importance of environmental cues and the profound importance of imitation. In fact, Skinner's work was profoundly important in advancing our understanding of environmental cues and imitation. Cues enable an organism to be sensitive to environmental features that "prompt" behavior to be emitted by organisms more likely to result in reinforcement. Thus, a student understands the importance of a teacher's entry into a classroom as a cue for good behavior that might result in an appropriate period of recess, just like circling buzzards might cue a hyena that carrion is available just over the ridge.

A less obvious, but perhaps more powerful, cue occurs when an organism observes the behavior of others, perceives the consequence of the behavior, and then uses that observed behavior as a cue to imitate. Imitation is a kind of vicarious operant conditioning and a near universal behavioral disposition among organisms with the biological capacity to perceive the behavior of others. Observe, imitate, and achieve a positive consequence – a new behavior is acquired. This cycle of imitation and cuing is significant in the nearly immediate process of shaping social behavior. Among the dozens of studies on imitation and cuing, decades of studies have demonstrated the power of imitated smiling behavior in infants as a means to initiate social and emotional contact. Other studies have demonstrated the enormous influence that infant imitation of speech sounds have on shaping language development, and the inevitable advance of language development as the environment and biology interact. Not until very recently did researchers describe the neural mechanisms of imitation, a topic taken up in more detail in Chapter 5. Suffice it to say that if Skinner had been witness to this discovery, he might have been given cause to reconsider his pure radical approach to the "black box." That reconsideration might have enabled him to view operant selection as the essential biological mechanism linking natural selection to behavioral expression. It is apparent that he was on the cusp of this transition with his "three kinds of selection by consequence." However, he did not live to push that concept beyond its very basic development.

Language and Culture

Language capacity is a particularly frequent and significant example of selection by consequences and biology interacting to produce an essential component of human social behavior and culture. One of Skinner's more noteworthy public discourses occurred over many years with Noam Chomsky regarding language acquisition. The debate underscores Skinner's determination to establish the profound importance of cue and imitation to behavioral acquisition – even the most "sophisticated" behavioral acquisition. Chomsky advocated for innate linguistic structures, while Skinner maintained that language was acquired through observation, imitation, and differential reinforcement of increasingly sophisticated language behavior.

Scholars across many disciplines now consider language to be enormously influenced by environmental factors (individual consequences and consequences maintained by social contingencies) while having a significant biological basis (natural selection). Steven Pinker (*The Language Instinct,* 1994) provides a seamless review of the range and depth in our understanding of human language between the extreme poles of blank slate and pure biological determinism. Regardless, scholars agree that language is among the most significant behavioral repertoires of culture that distinguishes humankind from other species. It is the operant (relatively ultimate cause) features of cuing, imitation, and operant shaping as biological predispositions in human behavioral response that bear so prominently on the emergence of all aspects of language development. Naturally selected auditory sequential processing and articulation mechanisms interact with operant shaping driven by individual and group consequences Skinner's three kinds of selection by consequences can easily represent the rich interaction that achieves language. Thus, human social behavior and culture are enormously advanced by a behavioral expression of this rich interaction.

Language enables humankind's significant advancement and flexibility in social behavior and cultural development both within and across human cultures. It facilitates the passage of accumulated cultural wisdom across generations and enables culture to have an emergent nature based on rapidly changing environmental events and diversity within individual perceptions. This human capacity for cultural plasticity is another characteristic of humans' distinctiveness and further evidence against reductionism and determinism. Obviously, language also provides us our most stable and powerful means to transmit the cultural units of humankind. Those cultural units have been identified and conceptualized most prominently as *memes*, by Richard Dawkins in *The Selfish Gene* (1976) and even more significantly conceptualized as *culturgens* by Charles Lumsden and E.O. Wilson in *Genes, Mind and Culture* (1981). Once again, it is of considerable importance to keep in mind Skinner's basic operant principles of cuing, imitation, and operant shaping as we think about Dawkins' concept of memes, and Lumsden's and Wilson's principles of gene-culture coevolution.

Units of Culture

Memes – a word play by Dawkins combining gene, mimic, and imitation – represent self-replicating units of culture and cultural evolution (based mostly on language). The term is used to refer to cultural elements, such as folklore, poems, songs, and manners, or more significant conceptual entities like religion or free will that are repeated and replicated throughout a culture. Dawkins offered as examples, "tunes, catch-phrases, clothes fashions, and ways of making pots or of building arches." He goes on to say that memes "propagate themselves in the meme pool by leaping from brain to brain via a process which ... can be called imitation."

Memes aren't always replicated accurately as they achieve exposure across a culture. In fact, they are modified, sometimes refined, or more often combined with other memes to form new memes – thus, providing a scaffold for cultural evolution. "Imitation, in the broad sense, is how memes *can* replicate. But just as not all genes that can replicate do so successfully, so some memes are more successful in the meme-pool than others. This is the analogue of natural selection". Yet, just as with operant shaping of behavior, the "natural selection" of memes is a far more rapidly occurring phenomenon.

Blackmore (1999) more recently suggested, "memetics (the study of memes) can provide a new explanation for the origins and evolution of the human brain. Since memes are, by definition, passed on by imitation, they must have first appeared when our ancestors became capable of imitation ... memes were a new replicator that started evolving in their own way and for their own replicative ends. Since then, meme-gene coevolution has produced the enormous human brain ... for the propagation of memes". For Blackmore, "the critical turning point in human evolution arrives when our ancestors acquired the ability to imitate. Imitation can be a 'good trick' from the gene's point of view because it reduces the cost of learning".

Imitation enables the brain to spread memes, both for useful purpose and for meaningless purpose. Neural circuitry responsible for imitation doesn't necessarily evaluate the result of imitation. Blackmore suggests we consider a new way of considering human intelligence: view the human brain as "a selective imitation device". Selecting good memes to imitate discriminates among varying levels of intelligence. The capacity to discriminate "good" memes presumably differentiates individuals and groups. Spreading memes responsive to selection pressures advances cultural development, while imitating memes that lead to counterproductive rituals or superstitions is useless.

Culturgens and Epigenesis

References to Skinner and Wilson in the meme literature are rare and gratuitously brief in describing the relevance of their work; most frequently, references are notably lacking altogether. It is as if the importance of imitation and gene-culture coevolution were conceptualized with the dawn of memetics and Dawkins' selfish gene concept. Is the primary source of meme formation, modification, and replication anything more than reinforcement of cued behavior that has been imitated? Is the passage of memes by imitation anything different than operant reinforcement of cued behavior? Is the organizational scheme we call culture anything more than the imitation of cued behavior (memes) that achieve significant positive selection by consequences in the cultural environment? Is culture anything more than imitation or vicarious learning (replication, modification, or recombination)? Is the relatively ultimate cause of imitation the evolved capacity of living organisms to be cued and operantly conditioned? And, most importantly, is the relatively proximate cause of all gene culture coevolution the biological selection of neural circuits that subserve imitating social and cultural behavior? This final and critical question is the one ignored in memetics that was so elegantly addressed by Lumsden and Wilson (1981). Additionally, renewed consideration of group selection in the larger context of multilevel selection generally diminishes the selfish gene and memetics. Indeed, multilevel selection, of which group selection is a subset, acknowledges the selfish gene as a *unit* of selection but refutes it as the only unit of selection (Sober and Wilson, 1998). In fact selfish genes, by definition, can't interact with their environment.

Genes, Mind, and Culture (Lumsden and Wilson, 1981) significantly advances the basic concept of memetics *and* includes description of their essential biological character – however, the book's terminology and mathematical modeling seems to have been effective in significantly limiting its broader appeal. It still remains a powerful foundation for scholars attempting to advance our understanding of social behavior and culture within the larger context of evolutionary biology. They describe epigenesis as the various "processes of interaction between genes and the environment that ultimately result in the distinctive anatomical, physiological, cognitive, and behavioral traits of an organism." Skinner would likely add that operant conditioning is the basic mechanism of that gene/culture interaction and would be in full agreement with Lumsden's and Wilson's teaching and learning continuum of simple learning, imitation, teaching, and reification. Skinner might have been inclined to substitute cuing and/or modeling for teaching, and clarify reification as conceptual representation via symbolic means.

The process of epigenesis leads to the emergence of epigenetic rules that achieve their most significant expression in the construction of culturgens, the term coined by Lumsden and Wilson (1981) to identify the basic units of human culture – transmissible behavior, mentifacts, and artifacts. The emerging regularities of epigenetic rules that canalize those traits, ultimately "occur at all stages of development, from protein assembly through the complex events of organ construction to learning" and serve "to keep culture on a leash." The undue criticism resulting from the "leash" imagery might be answered by using the popular rubber band image to represent the interaction of genes and environment in the emergence of intelligence. Epigenetic rules provide for human expression of culture but allow for significant flexibility in their bearing. The neural mechanism that provides the biological foundation for the rich interaction between environment and biology, gene/culture coevolution, and the emergence of epigenetic rules has eluded us until the very recent past with the discovery of mirror neurons in the early 1990s. Chapter 5 offers a more lengthy description of this mechanism as a means to biologically bridge operant conditioning, operant imitation, and the emergence of social behavior.

The Operant Basis of Sociobiology

The functional behavioral basis for Wilson's gene-culture coevolutionary interaction is provided by Skinner's operant paradigm – contingencies that select individual behavior and special contingencies maintained by the social environment. Epigenetic rules are apparent as we observe the increasing canalization in behavioral dispositions that emerge from the constant interaction among the operant shaping provided by the imperatives of natural environmental consequences, the developing contingencies of human social behavior and culture, and neural information processing modules. Certainly, Wilson and Skinner would agree that this rich gene-culture coevolution achieves its finest expression in the human mind. Epigenesis is dependent on operant shaping and our emerging mind is the ontogenic expression of epigenesis. A disciplined and wide-ranging study of epigenetic rules will guide our best understanding of human nature and its range of expression. Most elegantly stated,

> *A society that chooses to ignore the implications of the innate epigenetic rules will still navigate by them and at each moment of decision yield to their dictates by default. . .Economic policy, moral tenets, the practices of child rearing, and virtually every other social activity will continue to be guided by inner feelings whose origins are not examined. . .It will continue to live by the 'conscience' of its members and by "God's will. . ."*
> *The decomposition of social behavior into objective functional units, the discovery of new and sometimes surprising epigenetic rules, the measure of human genetic diversity, the tracing of the microevolution of the features of behavioral epigenesis, the paleobiological reconstruction of the origin of culture, the retrodiction of ethnography and connection of the results to new and more powerful covering laws in economics and sociology, perhaps even a sighting down the world-tube of possible future histories – these are the activities that will come increasingly to occupy the social sciences as the links between biology and the study of culture are more powerfully forged (Lumsden and Wilson, 1983, p. 358–362).*

There is full agreement between two of modern science's most significant intellects. Empirical investigation of the accumulated behavioral repertoire we describe as human nature will reveal our species' path to survival. Should we continue to accept human intuition based on folklore, myth, and anecdotal evidence in the face of empirical findings, we will continue our inevitable path to extinction. Should we choose to critically examine and judiciously apply the empirical findings of our best intellects, "Societies that know human nature in this way might well be more likely to agree on universal goals within the constraints of human nature" (Lumsden and

Wilson, 1981). In his typical eloquent fashion, Wilson (2006) captures the essence of the enormous challenge facing our human species – "a new and very odd species came shambling into our universe, a mix of Stone Age emotion, medieval self-image, and godlike technology. The combination makes the species unresponsive to the forces that count most for its own long-term survival."

A Conversation

What follows in Chapter 4 is the edited transcript of the conversation between B.F. Skinner and E.O. Wilson that occurred at the twilight of Skinner's life. Explanatory notes are included at the end of the chapter and are identified with numbered references throughout the text of the conversation. Occasional parenthetic comments are included within the body of the text to provide readers an implied detail or an assumed common basis of understanding. The comments are intended to enable readers to achieve a more seamless reading of the conversation.

The meeting was initially guided by the efforts of Skinner to establish the basis for the conversation. The two eminent thinkers then worked to achieve a frame of reference for their conversation and common vocabulary to explore their mutual interest – human nature. Although the conversation is less than successful in achieving a complete exploration of their shared interest, it promotes a serious consideration of their respective conceptual frameworks, is richly suggestive of their scientific linkage, and promotes serious reconsideration of Skinner's legacy. The conversation is also noteworthy in revealing their sense of responsibility to and passion for ideas in the pursuit of truth – a passion inspired by a desire to help save humankind from itself.

Chapter 4
Conversation Between B.F. Skinner and E.O. Wilson

(Harvard University, Cambridge, Massachusetts, November 19, 1987)

E.O. Wilson: This is going to be a conversation that I will have with B.F. Skinner. This is Ed Wilson. He invited me by to talk about sociobiology. Our relations have always been very friendly and I look forward to it. This should be an interesting talk this Thursday morning.
B.F. Skinner: We will start with a basic statement. I assume that you are what I call a behaviorist. You would accept that an organism is a biophysical and biochemical system, a product of evolution.
E.O. Wilson: I am.[1]
B.F. Skinner: (You would identify yourself as a behaviorist) . . . and nothing else?
E.O. Wilson: Yes – so far, so good.
B.F. Skinner: That would include not only genetic behavior, but also the kinds of behavior that can be learned because of genetic processes. Of course it (behavior) always goes back to genetics.

I would want to identify culture as something that has evolved in the human species. I think human cultures are quite different in kind from the social behavior of other species. Human social behavior may resemble the social behavior of other organisms, but it is also something quite different. That difference is due to the human vocal musculature and the operant control of that musculature.

I don't know how you feel about my claim that it was just the acquisition of operant control of the vocal musculature that made all the difference. Only in the human species can you reinforce sound production with the vocal cords and the vocal apparatus as an operant. It doesn't produce consequences (in other organisms). You can get a good imitation in the parrot, but it's very hard to get a parrot to make a particular sound pattern because of the consequences that follow. It is (simply) imitative and has many good reasons for being that. But if you take that to be the difference (my book on verbal behavior is based on that assumption), then development of cultures came about as ways in which the individual could profit from what has happened to other members of the group, and others long dead who are recorded in verbal behavior of one sort or another. If that is all there is . . .

Now, many people want some sort of (cognition) as a spirit of the time, and the whole cognitive revival of mentalism, as if it were something apart from biological

P. Naour, *E.O. Wilson and B.F. Skinner*, Developments in Primatology: Progress and Prospects, DOI 10.1007/978-0-387-89462-1_4,

activity of the organism. I reject all of that, and I more or less assume that you would.[2]

E.O. Wilson: I'm a behaviorist in the sense that I understand you to define it. That is, I hold a rigorous cause-and-effect objectivist view of behavior.[3]

B.F. Skinner: It is not merely cause and effect of the push-pull, stimulus-response sort, because selection comes in at all three levels (presumably, natural selection, individual operant selection, and group operant selection). And I think that is something very few people properly understand; that is, the difference between the causality we can observe in physics and the selection action on the behavior of living things. I think Ernst Mayr would be one of those who would want to use the word *organization* to define *living thing*. Would that be what you would think?

E.O. Wilson: Yes, that's right.[4]

B.F. Skinner: But I think that it (organization) suggests complexity. The organism is something involved in itself. But I think that it doesn't have the Darwinian substitute for purpose in it, which is selection. Would you feel that way? I'm sure he (Mayr) is a good evolutionary theorist, but it seems to me that he doesn't take advantage of that (complexity) to define biological structure as always a product of selection rather than just building – putting things together.

E.O. Wilson: I always thought of Ernst (Mayr) as an ardent selectionist and I tend to be over at that pole myself. I'm not quite sure that I follow you; what is he leaving out?

B.F. Skinner: Does organization imply selection to you?

E.O. Wilson: Organization is a product of selection.

B.F. Skinner: Yes, that's right; you can't get (organization) without having it emerge through variation in selection.

E.O. Wilson: Yes.

B.F. Skinner: But I don't think as such, I suppose a computer could be said to be highly organized and you could trace the history of computers to find out how (computer organization) evolved. But, it didn't evolve through the preservation of variations because of their consequences. Apart from that, the main thing on which I feel I differ from you is that sociobiology leaps a little too cavalierly from socio – to bio –. Sociobiology seems to leave me out. I'm in the middle.[5]

E.O. Wilson: How well aware I am of that. It certainly struck home with me when *Sociobiology* was reviewed in the *New York Review of Books* by (C.H.) Waddington (8/7/75).

B.F. Skinner: Oh, yes.

E.O. Wilson: Waddington said, "Fine, but he left out the mind." And I believe he was making the same claim (as you suggest of me in leaving you out of sociobiology).

B.F. Skinner: Yes, I think so. I think mind, person, or individual have to come into it.

E.O. Wilson: Yes.[6]

B.F. Skinner: I've been doing a series of papers on the relevance of what we can respond to within our own body and how it (our response) relates to the conditions that produced what is going on in the body. And so that sort of thing (mind, person, individual) comes into it. Natural selection gives you the organism, an organism that is so constructed that it does all kinds of things, including (emitting) its behavior.

And you have all of the extraordinary things (social behavior) that you have shown. Do you ever stop to wonder about the intervening steps in the evolution of complex behavioral patterns? Evolutionary steps at every level that would have to be selected by consequences. Now isn't that a real problem?[7]

E.O. Wilson: It (choice behavior and/or the evolution of social behavior and culture) is a very real problem. In fact, I got busy with Charles Lumsden, who's enough of a mathematician and theoretician to mathematically model choice behavior. We came up with two books: *Genes, Mind and Culture* and *Promethean Fire*, the latter being a popular version. We were attempting to follow those evolutionary steps through the various levels of organization, from the genome up to the final behavioral product. We certainly were aware that as far back as your paper, you were drawing parallels between selectivity and choice-making, whatever the agent of choice.

We recognize that with environmental circumstances through evolutionary time, you can have a genome evolve so that it prescribes ant-like behavior . . . although, even ants learn some things. You can prescribe it so that the organism will always respond the same way, like a male mosquito responding to a monotone sound of a certain frequency. Given the right stimulus, it responds only in one way; it will not respond at all if given a wrong stimulus.

B.F. Skinner: Yes.

E.O. Wilson: . . . Or the organism can advance so as to respond very strictly in fashion A to stimulus A1, B to B1 and so on. Then it is possible to have a genome that produces a nervous system that responds according to a very few rules of thumb. Imprinting is an example of that. You can modify your specific behavior (in response to stimulus B1), but this new response will tend to be what you could call an invariable. The object to which you respond and the exact fashion in which you respond is strictly dictated – usually during a very brief period in early life.

(This modification of behavior can continue) on to where the choice-making is vastly more flexible, and I suppose that is where biologists perceive an array of possibilities.

B.F. Skinner: Yes, go ahead.

E.O. Wilson: I was about to finish by saying that I see mentalists now largely confined to a few social scientists – ideologues and humanistic scholars having been put to flank. In other words, behaviorism as a philosophy is totally in control, thanks to its support from neurobiology and behavior study.

There might be enough difference in perception to create a certain amount of dissonance among people coming to this area from different disciplines. If not clearly articulated, the degree of restriction on choice-making behavior at the upper levels of human cognition leading to culture can be the source of (significant) misperception.

Where I have run into trouble most recently has been in taking a somewhat hard-line determinist position that the number of choices human beings are likely to make over a wide range of cultural environments is really relatively restricted, and that there is a tight interaction between the culture as created and (biological) predispositions.

I wonder if this is really all that distant from your own position.

B.F. Skinner: No.
E.O. Wilson: The assortment of primary reinforcers that are there tend to create a culture of a certain character.
B.F. Skinner: Oh, yes . . . right.
E.O. Wilson: Then the culture itself tends to feed back in satisfying and narrowing the choices for the genome to find an expression. And that doesn't sit well with a lot of people, even some biologists. But it's hard to come to grips with it in any objective discussion, because it's very difficult to measure. Charles Lumsden and I tried to devise a study measuring the baseline of cultural diversity given certain levels of restriction on the choice-making.

I think that is where we ran into most of our trouble. People felt we were making our models overly objective. There wasn't enough known to make our models that explicit. Furthermore, the baseline of cultural diversity is only the beginning. There are so many other things like historic accidents and remarkable leaps of creativity that throw that all out of kilter.
B.F. Skinner: I would point to those things if I would ask you to assure that when you define choice, you are not assuming that there is freedom exercised by the individual . . . there is some reason why you go right or left.

If you move back one step from the act of choice to the determining setting, when you are deciding to go or to not go, you haven't gone yet. *You are working on yourself* to produce an act (I don't like to use *choice* I prefer *decide*). If I go there, I'll get tired and so on . . . Or, I'll do this and so on . . . Of course, you have also learned all sorts of techniques, perhaps flip a coin or something of that kind to avoid the indecision. I recognize all that behavior of making a choice. It is something that occurs within you in addition to the experience that brought up the puzzle in the first place.[8]

I want to return to your mention of imprinting. I wonder if you are aware of an imprinting experiment done by Peterson in a laboratory in old Memorial Hall. He was able to operantly condition young ducklings to peck a key to make a box come closer. Key pecking by the ducklings was reinforced by an imprinted object (the box approaching). The only reason I think that its important is that it may be a clue regarding what happened when operant conditioning as a process evolved.

Your ants do all kinds of things (in response to specific stimuli). For example, the soldiers will respond to air puffs by coming out.

I don't think it's possible, but could we show that the ant that goes to a leaf and bites out a section could be made to press a tiny lever to have a leaf appear? In other words, is a leaf a reinforcer as well as the stimulus that elicits a genetic response?
E.O. Wilson: Yes.
B.F. Skinner: I think you can always show that it would be to the advantage of the species if those stimuli (leaf) which are now releasing behavior (biting) proved also to be reinforcers. When they serve as reinforcers, they are different in that they can reinforce a range of emitted behavior. Thus, other behavior could develop in the lifetime of the individual; that is operant conditioning.

I think we might be able to identify that the point at which such stimuli become reinforcing is a change in the individual and in its genes. (The organism's capacity to be operantly conditioned), supplements and strengthens the survival value of the (initial) reflexive or ethological behavior.[9]

That's another matter, I think . . .

E.O. Wilson: An extremely important one. You know, it's an interesting problem that you raise and I don't think that it's ever been explored. It's an example of how a particular viewpoint about the evolution of choice or reinforcement (operant conditioning) could lead to some explicit experiments. We are accustomed to addressing all these questions as far as social insects are concerned in the ethological manner. We see X number of behavioral acts performed and we see the circumstances (discriminative stimuli or cues) under which they are performed. We note that only a very narrow range of circumstances evoke the behavior.[10]

B.F. Skinner: Yes.

E.O. Wilson: And then we run down the behavior to find which of these signs, stimuli or releasers (discriminative stimuli) are involved. We know that there is a certain amount of learning going on in terms of orientation when the animals go out looking for this. But then we come up against the whole issue implied by the example of the duck acting to bring the box close by what used to be called appetitive behavior . . . and that is searching behavior. Searching for what?[11]

B.F. Skinner: . . . there is a certain purpose element in the word "search" that I don't think belongs there.

E.O. Wilson: Well, perhaps not. But there is no question that if you starve an ant colony, then the foraging workers begin to scout through the territory in larger and larger numbers and more intensively and they respond to lower amounts of the stimuli.

B.F. Skinner: Yes.

E.O. Wilson: It would be an extremely interesting experiment to see whether you could get an animal to do some auxiliary act in an operant manner (such as extend the range of its free operant behavior).

B.F. Skinner: Yes . . . you are thinking correctly – topography (or range of free operant behavior) is not part of the innate behavior.[12]

E.O. Wilson: Yes. In order to increase the stimulus that would lead to the goal.

B.F. Skinner: Yes.

E.O. Wilson: . . . or even replicate in some way the exploratory behavior. So, I don't know the answer to that. But you know, in abstract, one would expect that there would be an increasingly large program of auxiliary operant learning (behavior) of this sort that would tend to bring organisms to a relatively narrow set of accomplishments.[13]

B.F. Skinner: An ant is (operantly) wandering around and comes upon a trail and then follows the trail. Can we prove that the trail is a reinforcer? That would be the question, wouldn't it?

E.O. Wilson: That the trail would be (a reinforcer)? Yes, that's the question.

B.F. Skinner: The ant could stand up on his hind legs and there would be a trail, it would learn . . .

E.O. Wilson: Exactly. We want to know whether the ant would actually do something.

B.F. Skinner: That's right.
E.O. Wilson: Conditioned in an operant manner, in order to get to a trail.
B.F. Skinner: Yes exactly. Now why not – I've seen you lay down a trail. I know it can be done.
E.O. Wilson: Yes.
B.F. Skinner: Why not put an ant with a light to one side and the trail (to the other) and you always put the ant between the light and the trail so that it must go away from the light to get to the trail.
E.O. Wilson: Yes.
B.F. Skinner: And then show that it does that (the ant moves toward the trail) statistically more than it would do if there were no trail there.[14]
E.O. Wilson: Sounds like a good experiment.
B.F. Skinner: Yes, put a student on it (laughing). It could very well be that events are susceptible to operant reinforcement. But it (moving toward the trail) would be the right thing to have happened; it would have been very useful anyway.

I'd like to mention also that I have been very much influenced by something that happened (in my very early years as a researcher). I can give you the exact time and so on – the old pigeon project we were working on, training pigeons to guide missiles. And we had a lot of time on our hands because Washington couldn't make up its mind.

We did an experiment one day. We put a hungry pigeon in a box and we had a food dispenser and a switch that we could operate by hand. We put a ball in about the size of a ping pong ball. When the pigeon would look toward the ball we would reinforce looking. When it got closer we would reinforce, and in no time at all he was knocking that ball around as if he were a squash player.

Well, this was astonishing because I had never seen behavior shaped that way before. Now I've gone on and done all sorts of things. Shaped very complicated performers like the demonstration that I had for my students of pigeons playing ping pong. There is no limit to what you can shape by way of complex behavior.[15]

Now that (understanding the means by which operant shaping is accomplished) is what makes it very interesting for me to go back and speculate on what has happened to teach salmon to leap very high.

If you had a fish that doesn't leap, I could teach it (using operant shaping) to leap. I would (start by placing) a barrier just below the water it can slide over to get food. Then I would (gradually) raise it until it would fly over. It would go up and up and eventually you would have a fish that leaped like a salmon.

(Another example of operant shaping in the natural world is) ... the archer fish that comes to the surface and squirts water up and hits an insect hanging on a branch. I could hardly believe that happened, but there were photographs of it.
E.O. Wilson: Oh, there are even more astonishing things happening, but they often can only do that one thing.
B.F. Skinner: Yes, but can you imagine the intervening stages that would shape such behavior? How would I teach a child with a squirt gun to do that?
E.O. Wilson: Right.

B.F. Skinner: You must have reinforcing consequences at every stage. You just can't simply say oh, well, there's some pattern of genes and they just simply occurred.[16]
E.O. Wilson: Yes.
B.F. Skinner: It's too big a thing to occur at the level of (genetic) variation. And I keep looking at other examples. Take the kangaroo born as a little bit of a fetus, but the darn thing climbs slowly up its mother's belly, dives into the pouch and finds a teat.[17]
E.O. Wilson: Right.
B.F. Skinner: What the devil? What was the shape of the kangaroo when it didn't have too far to go?
E.O. Wilson: That's right.
B.F. Skinner: Apparently, Australia was a very nice place for all things to evolve in all sorts of directions. But I must say my belief in the lack of any divine principle is shrunken by these elaborate things. The Bowerbird, how the devil can you explain that kind of courtship dance? You can, I think (by invoking operant shaping). I would explain the rain dance pretty much as I would explain the courtship dance. For me, the selection would be operant (short-term shaping) instead of your kind of thing (a process of long-term natural selection).
E.O. Wilson: Right.
B.F. Skinner: But if it is true that rain occurring is a reinforcer and it reinforces anything that's going on at the moment, whether it really is functionally connected or not.
E.O. Wilson: Yes.
B.F. Skinner: Then inconspicuous things going on at the moment are more likely to be shaped by it. And the more conspicuous things you do at the end of a drought, the more things you do (in the rain dance ritual). The thing develops into a very elaborate kind of thing.
E.O. Wilson: Good point. You get an accretion (of unrelated behavior that are reinforced by the timely occurrence of rain).
B.F. Skinner: Right.
E.O. Wilson: Reinforced rituals that are performed.
B.F. Skinner: Yes. And the courtship dance might have been that. You attract attention with just a quick move and then it's more and more and more that does it.
E.O. Wilson: That's right.
B.F. Skinner: (This brings to mind) my experiment on superstition; I have done it many times. I was giving a lecture at the Royal Institution of London. I began it by putting a hungry pigeon in a box with a food dispenser operating every few seconds. And I covered it up (and announced), at the end of the lecture we will see that this pigeon will have learned what to do to make the food dispenser operate. And, sure enough, it was hitting the wall with one wing.
E.O. Wilson: That's remarkable.
B.F. Skinner: I have had pigeons learn to bow, to dance and so on, but just by accident. Within seconds, you see behavior being (superstitiously conditioned). (The bird is doing) something when the food appeared, so there is a tendency to do that

same thing again. If you hit (reinforce) it twice, then you are in (the organism has likely acquired the behavior).

E.O. Wilson: These questions about flexibility and the insertion of primary reinforcers are generally not the kind of questions that biologists raise. That is one reason they haven't been well explored. On the miraculous feats of kangaroo fetuses and archer fish spitting and so on, this is a favorite subject of the ethologist compared to zoologists. I have done a few of these myself. For example, the remarkable weaver ant. When the nest consisting of silk walls in the tree tops is torn open, out come worker ants carrying larvae that are at the mature stage and able to produce silk. The workers use them like silk shuttles to weave the nests back into place.

How on earth could that come about? Burt Höldobler and I have actually searched the world and come up with intermediate stages, including the most primitive that I found in South America. It was a species in which the larvae simply pay out a little bit of silk onto the wall of the nest and help reinforce it while they are spinning their own cocoon. Now that has a selection advantage and it shows you how, in terms of genetic prescription, you don't have to take the leap to the weaver ant stage, but you can go a little stage by little stage. That's standard reasoning. But what seems to me to be left unanswered – and it's the grandmother of questions as far as I'm concerned – when you get up to the origin of culture, how many primary reinforcers are there and in how many ways can they be utilized? And, by what range of stimuli?

You know, in terms of the most primitive forms of programmed behavior, a good engineer now can devise a cerebral mechanism that goes through the same steps as a kangaroo fetus.

B.F. Skinner: Oh, yes.

E.O. Wilson: At that point, the neurobiologist and the animal behaviorist tend to stop and say, why should we worry about all intervening stages of reinforcement? All we have to do is have an automaton that moves to the right when it receives a stimulus; it moves to the right until it receives additional stimulus; then it moves forward and so on until the act is complete. But it becomes much more interesting when flexibility is introduced so that new things can be added to the repertoire or the organism can make choices among the sources of the reinforcing stimuli.

B.F. Skinner: Al Knoll who was here to give the William James lectures, said he and Herb Simon have computer systems now worked out that do learn operant behavior. That is, the consequences changed them so that they behave in a given way. This is perfectly feasible. After all, here we do this (presumably sit and engage in a verbal behavior). I don't see why a machine can't be made to do it.

E.O. Wilson: That's right.

B.F. Skinner: It will be a different machine and it won't help tell us very much about what was going on in the nervous system.

E.O. Wilson: It can be proved.

B.F. Skinner: I'm a "black box" man myself. I think I was the first behaviorist to say I don't care about what's going on in the mind and I don't care what's going on in the nervous system.

I suppose I got it from Jacques Loeb (*Comparative Physiology of the Brain and Comparative Psychology*) and through William Crozier (former Department Chair

of Physiology at Harvard). I read Loeb in college and was lucky enough to get Crozier here.
E.O. Wilson: I knew him slightly myself when I was a graduate student.[18]
B.F. Skinner: Did you?
E.O. Wilson: Yes.
B.F. Skinner: He gave me five years of support – apparatus space and so on. I learned a little something being in biology rather than in psychology. I think it was all a really good thing for me. But I think he (Crozier) just passed on. Loeb is said to have resented the nervous system.
E.O. Wilson: I think if Loeb had all the armaments of neurobiology at his disposal, or could see others using it, he may have gladly extended his mechanistic approach down into the nervous system.
B.F. Skinner: Well of course, he had nothing but tropisms, reflexes, stimulus/response, and cause and effect sorts of things. As far as I know, he didn't have the notion of selection at all.
E.O. Wilson: Right.
B.F. Skinner: But he was in close correspondence with Ernst Mach. And it was Mach's science of mechanics (*The Science of Mechanics*, 1915) that had a big influence on me. The whole idea of science emerging from technology, the slow shaping of the behavior of the scientist by his consequences. It's a beautiful example.
E.O. Wilson: And minimalist.
B.F. Skinner: Just like the kangaroo fetus learning to climb.
E.O. Wilson: The minimalist explanation of science as the maximally efficient way of describing nature always appeals to me.
B.F. Skinner: Yes. I think the brain scientists are going to tell us eventually what's going on inside. They are much farther away from it than they seem to believe ... and I don't think introspection will ever tell us. No current psychologist, to my knowledge, practices introspection. Possibly Freudians and William James with his stream of consciousness (were the last). Nobody does it any more for very good reason. First thing, you don't have sensory nerves going to the parts of the brain that are doing all the important things. And, you can never agree on how you use the terms, because the person that's teaching you doesn't know what you are talking about. For three thousand years people have been trying to do it. It just can't be done. Let the brain scientists do it. And tell the rest of the story.

There are gaps in my account too. I reinforce today and something (different) happens tomorrow. I don't know what happens in between, of course. There are other stimuli and responses (occurring in the natural environment). They are not the same thing (as occurred on the prior day's occasion), but somebody has to account for how that happens.[19]
E.O. Wilson: The first stage in any science is phenomenology. And then one dissects the machinery of it. Then if you are in evolutionary biology, you try to make sense of why it occurred.
B.F. Skinner: All right.
E.O. Wilson: In terms of natural selection, it's extremely hard it put in clear language. Somehow, I would like to see and understand from neurobiology and

psychology a clearer (research) program and formulation of the notion of reinforcement and flexibility. That, combined with neurobiological models that would help to explain how on earth this remarkable vocal ability and semantic memory evolved, would be for the delectation of evolutionary biologists like me. Remember, I am only an evolutionary biologist who is trying to reconstruct social behavior and without great concern about the exact form of machinery. What we would like to see is what you just alluded to – why culture evolved, exactly what culture is as a biological phenomenon, and why it only happened once. If it is such a great thing, why did it only happen once? This is a compelling evolutionary problem.

The evolutionary biologist tends to suspect that down in the machinery, there may be a simpler array of primary reinforcers than perhaps many scholars in the social sciences and humanities are willing to acknowledge. You know, they tend to be almost infinitely pluralistic.

B.F. Skinner: Yes.

E.O. Wilson: Imagine the concept that there may be a limited number of neurological devices and reinforcement mechanisms that are the primary source of motivation and behavior – the whole thing may be more constricted and simpler. That (concept) tends to repel many social scientists and humanities scholars from sociobiology or biology generally. Sociobiologists are one of the standard bearers of evolutionary biology and, who knows, behaviorism as well.

B.F. Skinner: Of course, why did (primary) reinforcers become reinforcing? The event is clear. Food in the mouth.

E.O. Wilson: That is the question.

B.F. Skinner: Sexual contacts and so on.

E.O. Wilson: Exactly.

B.F. Skinner: When did they (primary reinforcers) become (reinforcing)? My point is that they were (operative) during the first part of the ethology (behavior repertoire) of the species as elicitors and releasers of behavior. (Concurrently), when they became slightly reinforcing, that made it possible for new forms of behavior to appear in the life of the individual.[20]

E.O. Wilson: Right. That's a good sociobiological statement, that there is a bottom line. But there are a number of steps . . .

B.F. Skinner: Oh, absolutely. Well I never doubted that everything goes back to genes. But just how you get back to it is the problem.

E.O. Wilson: Well, let me just mention an idea that I know Waddington pressed in developmental genetics. I believe this is what you had been arguing as well – that is, the notion of change or the development of capacity for change and progress. Progress, if you want to define it, in higher survival ability.

When reinforcing mechanisms are built in by the genotype, extension into new forms of behavior is possible. If those new behavior are reinforced, they are incorporated into the repertoire of the organism. This opens a potential for new forms of behavior and/or an enlarged (behavioral) flexibility. It can be that this capacity for being reinforced by (the consequences of) some new form of behavior in an operant manner can lead to a new pattern of behavior. (This new pattern of behavior) produces the same result – availability of the primary reinforcer. But it also can result in a new behavioral act added to the old (repertoire). In other words, an expansion of the repertoire which we would then call behavioral flexibility, I suppose.

Waddington's original notion of genetic assimilation suggested that if there are a variety of genotypes in a population, some resulting in the capacity to explore and to be reinforced by new forms of stimuli that (exploratory) capacity may actually result in reinforcement occurring more reliably. That genotype would tend to survive and replace the older genotypes. Indeed, (the organism) might even have the propensity to explore like the pigeon bringing the box to you or the ant worker that wanders far away from the order trail.

B.F. Skinner: But as soon as the process of operant conditioning evolved, then there would be an advantage to the species if a lot of other behavior emerged which had no biological advantage. But if they were operantly conditioning . . . (they would likely emerge as common behavioral repertoire among the members of the species having inherited the same behavioral predisposition).

(Examples that come to mind include) the squirminess of the baby, random movements of a newborn organism that result in it locating the teat of a mother. That is one kind of (behavior) and you can understand why it got there. (However), it's just exploring the world around you and there was no parallel in the genetic behavior of the individual.

E.O. Wilson: Yes.

B.F. Skinner: I think quite clearly (another example is) the human face and vocal behavior particularly. You have to assume that a variety of noises became part of the genetic endowment because there was a culture waiting to pick them up and reinforce them as speech sounds. The only genetic contribution (providing a) behavioral predisposition is bringing the vocal musculature under operant control.

Subsequent evolution of those individuals who made many different sounds and could respond (more effectively) to a verbal community would be more likely to become part of (the community) and survive. Of course the sounds that get selected are (varied significantly by region).

E.O. Wilson: Yes, some (language sounds) that we can't even duplicate if we train ourselves (after an early critical developmental period). It's all very strange, but I don't think it is fundamentally troublesome.

B.F. Skinner: Well, let me advise the human sociobiologist – I have your paper on the subject. I think you have said it, but what do you think people who have this interest in the evolution of human behavior are most leaving out, particularly with special reference to biological underpinnings? What do you think they should be doing?

E.O. Wilson: Well, I would like to see more attention to reinforcing consequences and natural-selection consequences. (I would also like to see) recognition that it is just possible that there will never be any correction for the evolution of cultures. As I see it, natural selection was substantially responsible (for basic cultural structures) and then you are out (of alternatives). Then that problem was solved by the evolution of operant conditioning, which made it possible for the individual to acquire additional behavior that couldn't be prepared for.

B.F. Skinner: But of course, it, too, is only preparing for a future like the past. So, your behavior is shaped by contingencies and it will work when those contingencies prevail. Moreover, an individual couldn't learn very much through operant conditioning in one lifetime. But that was changed by the evolution of vocal operant control and the evolution of language, which of course enables the individual to profit from the behavior of others and what has happened to others. Then you get to the point where cultures prepare only for a world like that (which did the) selecting in past. Then we are stuck with governments, religions, and capitalistic systems (that are more responsive to environments of our evolutionary past).[21]
E.O. Wilson: Absolutely, you were getting close to the nitty-gritty and I wanted to make sure we were getting that right. We were talking about this evolutionary trap. That culture evolves upon a base of reinforcers which is in response to a primitive past and upon a base of a genotype – victories of the past. Our cultural practices evolved in a world in which the air was perfectly clean, the water was pure, and there were all sorts of land available for new crops.
B.F. Skinner: Right.
E.O. Wilson: And the ozone hadn't broken out with holes in it. And if you went to war, there was a good chance that a few of your men would be spared.
B.F. Skinner: Yes. Now what on earth are you going to do about this if the environment is changing very drastically and these cultural practices are stuck? Governments have their own immediate consequences to worry about. You can't get senators to promote one-child families or anything like that. None of this kind of –

You can't expect General Motors to manufacture automobiles that would go many miles to the gallon as they only go a few miles per hour. You can't do that. But you could say, oh well – there are millions of (possible) worlds and this one is going to go out. Another world will (will take its place) that is going to correct for this limitation at the third level.

Well, I think it could happen here, as well as anywhere else. I've just written a new preface for (the Penguin Edition of) *Beyond Freedom and Dignity* (1988) and I make the point that maybe a true third world estate is emerging. Scholars, scientists, teachers and the media who are talking about the future because they don't have immediate consequences to worry about if they are not paid by government, religions or capitalists. They are already beginning to have some effect. This is the third world, the world of the future (not the typical current usage of "third world") in which the fate of the earth (hangs in balance to be rescued by) friends of the earth.
E.O. Wilson: The rain forest campaign is an example of this. A quick awakening, a remarkably rapid awakening of the worldwide plight of the rain forest.
B.F. Skinner: Yes. Good heavens, is anybody stopping the cutting down of the rain forest?
E.O. Wilson: It's beginning to slow here and there.
B.F. Skinner: Yes. That's a good example.
E.O. Wilson: But it's coming into our ethics.
B.F. Skinner: Yes. We evolved with mechanisms taking care of the oxygen supply and now we are cutting them out. So, I don't know the answer to this, but I would like

to believe that a culture is going to evolve, quite apart from governments, religions' and capitol that are defined in terms of their mechanisms.
E.O. Wilson: Yes.
B.F. Skinner: Governments use aversive control (negative reinforcement) and capitol uses wealth (positive reinforcement). I don't know what religion uses. But you can't get religious people to concern themselves about the future of this world anyway. But, is it going to be a third step that will correct for the evolution of cultures (that is attentive to the) corrective (power) of operant conditioning and (effectively utilizing) operant conditioning (to be) corrective for culture?
E.O. Wilson: For pure genetic prescription?
B.F. Skinner: Right.
E.O. Wilson: And the inability to alter the way you do things except through natural selection over generations or more?
B.F. Skinner: And you seem to have been suggesting there are restrictions on how many different ways you can turn it this time.
E.O. Wilson: Yes.
B.F. Skinner: All we have is the product of fixed processes. And that is where things are.
E.O. Wilson: That's right. That's the key question. Let me see if I can make a little generalization that would align your basic suggestions about the need for a rational design of culture with the more biological approach, which tends to say there is a very limited way in which we can make that design.
B.F. Skinner: Yes.
E.O. Wilson: I'm inclined to think that what is sometimes referred to as the extreme behaviorist position is that we could design our culture and we can design it any way we want. It is just a matter of playing around, giving the right stimuli at the right moment. I know it's not fundamentally your position, but it is often thought to be by biologists. It is just a matter of designing what we want and it is done in kindergarten.
B.F. Skinner: Yes.
E.O. Wilson: The biologist response to that extreme position, which you know I don't believe you hold, is that we are infinitely plastic. We can't go that way because there are certain things that human beings are; they are programmed and cannot be easily altered – if they are altered the results could be shattering. A bad effect in any other realm.

It seems to me that the reconciliation of these two views (behavioral and biological) under a single umbrella is something with which I fundamentally agree. And the alignment of these two positions would come from the understanding that you can't do it genetically. You are not going to do genetic engineering (to accomplish it). We need to recognize that we have to start doing more design and planning of our culture, (a position) where I depart radically from Von Hyack and many of the other invisible-hand social philosophers. We do agree that we've just got to start, that the global condition of human population is in such a condition that we really have to design things for the long term.

You know, to eliminate aggression, we don't half understand anyway, you can't blindly go in the matter of Kampuchea and reorganize a whole society and say, adjust to that.

B.F. Skinner: No.

E.O. Wilson: Somehow we have to begin. Human science is a more aggressive exploration of what the primary reinforcers are, what their neurobiological bases are, and what their evolutionary history was. (We need) to find out where the maximum flexibility in domains of behavior was – others have minimum flexibility. Somehow there has got to be a kind of an intervening behavioral engineering that takes into account both biological constraints, which are due to our genetic history, and the potential that is afforded us.

B.F. Skinner: Yes. Did you use biological constrains as different from genetic constraints?

E.O. Wilson: No, I used them identically.

B.F. Skinner: Yes. I thought so.

E.O. Wilson: I'm talking to a general audience.

B.F. Skinner: Yes. We do design genetically; we have always done that with selection, like breeding cattle. And now we are designing with genes to some extent. But we have always designed personal repertoires by using education; you introduce variations and you also change the contingencies of selection. It has very seldom in any way affected cultures by changing the contingencies of selection. But of course we have always introduced new practices. We have introduced variations even though we can't affect the contingencies very well. Now what we have to do is change the contingencies of selection. Is that right?

E.O. Wilson: That's right. And in order to do that it seems to me we have to understand a good deal more about the developmental cognitive process.

B.F. Skinner: The power is in the hands of governments, religions and capitalism. That's the devil of it.

E.O.Wilson: Right. That's the problem all right. And it brings us back to the need for more science in decision-making.

B.F. Skinner: But it's got to be the people, because the only possibility is to turn to the people governed, employed or preached to and get them to change. Because the scientists are not big enough and they can't refuse to design better automobiles.

E.O. Wilson: And when all is said and done, most scientists are looked upon with great suspicion. Their philosophies are viewed as outside the pale.

B.F. Skinner: Especially behavioral scientists. Manipulating people like marionettes.

E.O. Wilson: Absolutely. Or for that matter evolutionary biologists including sociobiologists who are looked upon as potential monsters who want to tamper with the genes.

B.F. Skinner: Oh, yes.

E.O. Wilson: Or declare certain desirable areas of behavior out of bounds. But, it's true that we (society generally) make these fundamental decisions on how to manipulate society, but they are made by the people with the least training. Last year I was invited by the Roman Catholic Bishop to be one of four scientists (others were

Freeman Dyson and Roger Sperry) to talk with him at a special three day conference. It was so clear that the Catholic hierarchy is trying to make fundamental decisions about design of society. That's their business.
B.F. Skinner: Yes, they are.
E.O. Wilson: And the gulf between them and the scientists was just simply enormous. I gave them what I thought; Sperry and Dyson did an awful lot of waffling. But I gave them straight talk. Dyson is honest, but he was saying things I think they enjoyed hearing about the mysteries of the cosmos.
B.F. Skinner: Oh, God, yes.
E.O. Wilson: I was giving them straight materialism. But the gulf was enormous to get to their Catholic philosophers in attendance, (even though) they were trying to bring science back within the focus. But the point to be made, here, is that these are the people who are designing our society. Maybe the church is less potent than it was.
B.F. Skinner: Oh, no.
E.O. Wilson: But their equivalence in secular politics ...
B.F. Skinner: Fundamentalists ...
E.O. Wilson: Robertson, you know, with no concept whatsoever ...

So scientists should be made to feel a little more responsible. But how many scientists do you know who both have the inclination and the ability to go outside of their intensely competitive disciplines just to get funding and tenure – scientists who have the inclination to do it and who are going to be listened to when they begin to do it. It's not an encouraging scenario.
B.F. Skinner: The church is in terrible box right now of course because too many obviously good things are being proposed that are in conflict with the Bible and the old patterns. And they are losing the control of many of their people – for example, the number of Catholics who practice birth control.
E.O. Wilson: Yes. Well, I am somewhat optimistic, because there's a middle layer of science critics and science interpreters and responsible public figures who are able to collect this knowledge and in a deliberate manner put it into a form that allows it to be employed in decision-making processes. A very good example is Daniel Callahan's book called *Setting Limits* (1987). He talks about something you are very interested in – the aged. The whole dilemma society is in of life extension.
B.F. Skinner: Yes.
E.O. Wilson: And the exponentially rising costs and investment of energy of life extension. He is head of The Hastings Center which deals with bioethics. (Callahan) examines this area (of policy) in an objective, cool-headed way and comes up with the conclusion that there is a limit that we ought to agree to set, beyond which we do not use heroic measures or invest immense sums of public moneys in research. He suggests that the money and efforts thus saved should go back into the improvement of quality of life for older people.

But you know when all these questions finally come down into focus or into a conflict situation that compels focus, they are always settled on what are called moral grounds.
B.F. Skinner: Yes.

E.O. Wilson: And that leads us then to the whole issue of the relation of science and objective knowledge to morality. I happen to believe that we really can and should design a more explicit and objectively based morality. I have written a lot on this and gotten into a lot of trouble on it too.

I think it's exceedingly dangerous to develop all this scientific knowledge and then turn to a scientifically untrained group and say – you philosophers, or you theologians, or you supreme court justices are the repositories of our moral standards, use this knowledge the way you feel. I think that's a grave error.

B.F. Skinner: Yes, they appropriate the notion of the good.

I think we can say what we mean by what is good for the species, obviously what promotes the most genes. I think that's fair.

E.O. Wilson: Yes.

B.F. Skinner: What's good for the individual is what produces effective behavior and reinforces the development of a personal behavioral repertoire. What is good for the culture would be what makes the group that observes a set of practices survive – not only in competition with other groups, but by solving its own problems. What good would a culture be if there was only one culture in the world you could still speak of?

And so you've got three kinds of good there. And they are quite different. And the main problem is that the good for the culture isn't the reinforcer at all.[22]

E.O. Wilson: That's right.

B.F. Skinner: It would be fine if you could make it so. Now, I'm preparing for attack. What about altruism? Aren't there three kinds there again? They ought to be distinguished. Certainly one example is the male insect that copulates and dies or allows himself to be eaten by the female – he does something for the good of the species. That certainly is not for his own good.

Another example is the individual who can be conditioned to do something to another in return for something. Now that sometimes wouldn't be called altruism. You have to do it for no reason. But this is like the Christian notion of grace which is not supposed to get into heaven. If you do, you don't deserve to be in heaven. But, there is that whole thing there, too, of doing something which benefits others at the sacrifice to yourself. But I think you would have to show that there must be reinforcing consequences or the behavior wouldn't occur.

(Finally), with the level of culture, culture is out to get the individual. Because it's going to stop the selfish behavior of the individual, so the culture itself will survive. And the culture which was good enough to shame people or punish people for their enormous misuse of the environment would be more likely to survive. But the whole process of operant reinforcement isn't going to do that for you.

E.O. Wilson: Yes.

B.F. Skinner: The culture has got to supply the conflicting contingencies and somehow or other, punish profligacy and maintain some kind of austerity. These, I think, are moral issues.

E.O. Wilson: And do it perhaps by appealing to long-term consequences as opposed to short-term consequences.

B.F. Skinner: Yes.

E.O. Wilson: That's where the scientific education comes in.
B.F. Skinner: Yes. But how can you get people to take the long term into account? That's the thing, what you can do is let people give advice, but if you do that, horrible things are going to happen. It especially doesn't work if the horrible things are lying in the distant future because you've had no chance to find out whether the prediction of the distant future has been correct or not.
E.O. Wilson: That's right. Yes.
B.F. Skinner: If someone says if you drive that car with the brakes that way you will have an accident, you may take that advice but if you say if you drive that car on Sundays, there won't be enough gasoline for your great, great, great grandchildren, they are not going to do anything like that.
E.O. Wilson: Yes. Right.
B.F. Skinner: What you can do instead is to contrive current consequences that have the effect the remote ones would have if they were acting now.
E.O. Wilson: That's right.
B.F. Skinner: And that is where, for the state, the scholars, scientists and so on have got –
E.O. Wilson: Or sometimes protest groups.
B.F. Skinner: – have got to design a world in which you can live quite happily without consuming these vast quantities of things. I offer *Walden Two* as an example of that. I wrote *Walden Two* for a set of reasons which are no longer in existence. *Walden Two* is the ideal non-consuming, nonpolluting maximally socializing institution. And it's an experiment and therefore a variation which might survive if it has a chance to survive, because it is a much more efficient use of what it takes to live a happy life that is sustainable into the future.
E.O. Wilson: And will survive if it's protected from external aggression.
B.F. Skinner: That's the trouble, you see.
E.O. Wilson: It doesn't have a defense force.
B.F. Skinner: But, I think if we are going to design a better way, we need to experiment with it. Try it out.
E.O. Wilson: Right.
B.F. Skinner: And something like communities would be one way to do that.
E.O. Wilson: Yes. That's true. And that is being done to some extent.
B.F. Skinner: You were looking at your watch. – any particular reason to get back or what –
E.O. Wilson: No, no.
B.F. Skinner: I probably, I've probably talked myself out, I'm afraid.
E.O. Wilson: No. Well, as a matter of fact, I probably should, because I'm chairman of my department this year.
B.F. Skinner: Oh, oh.
E.O. Wilson: I have a department meeting later. So perhaps I should get on back. I thought we covered a lot of the ground that I had visualized in my mind.
B.F. Skinner: Yes.
E.O. Wilson: And I'll send you a copy of this tape.
B.F. Skinner: I think it would be worth having.

E.O. Wilson: I am going to give a good deal of thought to these issues that we are talking about, because I started giving a course in human sociobiology. I gave it last term.
B.F. Skinner: Yes, I remember it.
E.O. Wilson: It was very successful and I'm giving it again this term. And I find that I'm falling behind on the subject because I'm just finishing a huge book on ants.
B.F. Skinner: Oh.
E.O. Wilson: And so I know everything about ants but I have forgotten or haven't kept up with a lot of what's been going on in human sociobiology and all the related subjects. So I have in mind to get reacquainted and possibly even trying a textbook on human sociobiology. I think what we most need now is a textbook that addresses the various topics we touched on today. Many people have told me that. They say that if you want to make any progress out of evolutionary biology, then you have to have a textbook that explains it clearly to people in psychology and the social sciences. They don't want to have to go through all the journals and all the critiques and counter. They need a textbook – an honest textbook.

The next thing that you must do, as you have been arguing, is address the issues that are of major concern to psychologists having to do with development. It is very difficult to get these things lined up, but I think we accomplished a lot today – at least for me.
B.F. Skinner: Oh, yes. I think so.
E.O. Wilson: So I'd like to go away, reflect and maybe call on you again one of these days.
B.F. Skinner: Well, I wish Harvard professors did more of this.
E.O. Wilson: I think it's great.
B.F. Skinner: Isn't it strange that we don't?
E.O. Wilson: Yes. I think it's great. Several years ago, there was a group put together a celebration for Willard Quine.
B.F. Skinner: Oh, yes. I was in on that.
E.O. Wilson: Yes. And, well the thing was that I was asked to join a discussion group with Quine on ethical philosophy. So I went in expecting to see maybe a half dozen faculty members and some graduate students sitting around. It turned out to be Quine and myself and a small group that listened to us talk. Well, I was very uncomfortable I must say. And it's too bad that we don't have more of that with small groups of students.
B.F. Skinner: Yes – the old idea of a colloquium.
E.O. Wilson: Yes.
B.F. Skinner: People really sat down and all of them (felt a responsibility to the conversation). That was very good.
E.O. Wilson: That's great. I hope maybe when I can break clear a little bit, stop saving rain forests, get through with my book on ants, and get through this department chairmanship that I will have time to try some more activities like that. Are you going to the American Humanist Association Congress, by any chance? You may have just gotten an invitation.
B.F. Skinner: Oh, I hadn't planned to go to that.

E.O. Wilson: Well, I was asking, I don't plan to go myself. Although there's this academy of humanism and I keep wondering about whether there's something wrong with my reinforcers.
B.F. Skinner: Yes. I regard myself as a humanist in the sense that I am primarily concerned with human beings and what they do. But, Corliss Lamont, an honored humanist, says you've got to believe in free will. And I don't know whether I am a humanist if that is the case.
E.O. Wilson: Yes. That's right. That's the problem, of course. That, in order to be a constructive force, people who are collected loosely under the rubric of humanism need to have some organized position from time to time. But the more they organize the more manifestos they produce and so on. And the more like a secular religion they become and the more dissidents you are going to have. As an organization, I'm not the quite sure about its future – humanist societies tend to be unusually pallid.
B.F. Skinner: I think they ordinarily are pallid, that's exactly the word for it.
E.O. Wilson: Well, maybe humanism, as we think of as the force in human affairs, is going to emerge in a different form and that's just through the accretion of a body of knowledge.
B.F. Skinner: Yes. You want to make sure humanism is not ever considered a religion because that would require many changes; a lot of textbooks would have to be changed to let all the other (religious) stuff in.
E.O. Wilson: That's true. We came dangerously close.
B.F. Skinner: But what is it if it isn't (a religion-like system)?
E.O. Wilson: We came dangerously close I think with this last movement in Alabama to have it called a religion.
B.F. Skinner: Oh, yes.
E.O. Wilson: Then it could become equivalent to Southern Baptist. Well, I'm going to run along.
B.F. Skinner: It was delightful.
E.O. Wilson: It really was tremendous.

Notes

1 *I assume that you are what I call a behaviorist.*

Behaviorist, behaviorism, and behavioral: The use of these terms throughout the text is a fundamental element to adequately understand Skinner's position on the array of topics he addressed with Wilson. Equally important to the conversation at this point is Wilson's acknowledgement that Skinner's characterization of him as a behaviorist is correct.

John Watson was first to use the term "behaviorist," as Skinner intends it to be used, when he coined it for the title of his 1913 article published in Psychological Review – "Psychology as the Behaviorist Views It." This usage obviously predates Skinner's conceptualization of operant conditioning with his publication in 1938 of *The Behavior of Organisms*. Overt and quantifiable behavior that is readily open to empirical manipulation is the foundation of behavioral psychology and is of primary importance to a biologist with serious interest in ethology and the evolution of social behavior – precise observation and careful quantification are the rule.

Interestingly, these same behavioral techniques have been more recently utilized to great advantage by the behavioral neuroscience community to reveal the "black box" during real time human cognitive processing. For example, technology has enabled researchers to synchronize gathering highly sophisticated event-related potential data or fMRI data with observable and measurable behavioral task performance. Such is also the case for researchers exploring the mechanisms of echo neurons and mirror neurons.

2 *... Human social behavior may resemble the social behavior of other organisms, but it is also something quite different. That difference is due to the human vocal musculature and the operant control of that vocal musculature.*

Humans became increasingly social when vocal musculature came under operant control. Skinner's *Verbal Behavior* (1957) was a comprehensive explication of this theory – a theory he revisited on many occasions, including this conversation. "By behaving verbally people cooperate more successfully in common ventures." (Skinner, 1981). Skinner made claim that all other aspects of human social uniqueness and culture are reducible to this human capacity for the operant control of the vocal musculature.

> *Now, many people want some sort of (cognition) as a spirit of the time, and the whole cognitive revival of mentalism, as if it were something apart from biological activity of the organism. I reject all of that, and I more or less assume that you would.*

To be sure, Skinner hyperbolically reinforced this same sentiment in his final public appearance before the 98th Annual Convention of the APA during his extemporaneous comments following his receipt of the APA's lifetime award. "So far as I'm concerned, cognitive science is the creationism of psychology." The statement also appears in his posthumously published article that revisited his APA presentation (Skinner, 1990).

3 *... That is, I hold a rigorous cause-and-effect objectivist view of behavior.*

Wilson's usage of "cause and effect objectivist" was rather flatly stated, with no real elaboration offered. Points of distinction he offered later in the conversation make it apparent that he was keenly aware of the charges leveled at him for reductionism and determinism. He was also clear about his usage – by "cause and effect objectivist," Wilson referred to mechanistic attributes of behavior that lack the range of behavior of more complex and/or organized behavioral dispositions or behavioral repertoires. According to Wilson, the cognitive elements included among social or intelligent behavior are not mechanistic, but emergent materialist elements of highly sophisticated social behavior patterns that arise during gene-culture coevolution.

4 *... I think Ernst Mayr would be one of those who would want to use the word organization to define living thing.*

Organization, complexity, and *advancement* are concepts that occur frequently in the literature of evolutionary biology and evolutionary psychology. However, their usage in discussions regarding the evolution of behavioral repertoires frequently achieves more misrepresentation because their context often lacks completeness or clarity. Additionally, their usage can be fraught with conceptual misrepresentation and laced with implied values. The terms should not be read to imply ascendant progression or direction. In that regard, *organization* is the least troublesome of the terms and tends to connote the least implication of progression or directionality.

Skinner's reference to "selection comes in at all three levels," without elaboration bears significant explanation. His presumed three levels in the comment can each be described in behavioral psychology and are first explicitly introduced in "Selection by Consequences"(1981), and then significantly elaborated in his final publication ("Can Psychology Be a Science of Mind?), in which he suggests that the behavior of the organism is the product of three types of variation and selection. (1) Darwinian natural selection of innate, automatic, and stereotypical survival behavior that are released by an environmental stimulus, (2) individual behavior learned by individuals as a result of operant shaping in response to a particular environment but that are not stable enough to play any part in evolution, and (3) social behaviors that are modeled, primed (cued), imitated, and then performed to achieve reinforcement most often available via group contingencies and maintained by the social environment. Skinner claimed these social behavior achieve common expression as an aggregate of common group behavior we call culture. See Chapters 1 and 3 for more elaboration of Skinner's trio of selection by consequences concept.

Skinner goes on to say in his final article that study of "the evolution of a culture is also primarily a matter of inferences from history," and claimed that "only operant conditioning occurs quickly enough to be observed (in the laboratory) from beginning to end." In fact, he goes on to claim that "the role of variation and selection in the behavior of the individual is often simply ignored. Sociobiology, for example, leaps from socio- to bio-, passing over the linking individual."

Skinner's radical position left little room for Wilson to suggest additional considerations that might have bearing on their unfolding conversation. Wilson's frame of reference might have moved him to suggest that Skinner's three part variation and selection reference might be further illuminated by considering that in each successive case, behavior is increasingly "organized" over evolutionary history. The interaction of environment and genes achieves expression in neural tissue as a coevolutionary process. That process is anything but mechanistic in its outcome because of the increasing malleability of neural tissue as we travel the phylogenic continuum. Much more on this complex topic can be found in Chapter 3.

5 *But I don't think as such, I suppose a computer could be said to be highly organized and you could trace the history of computers to find out how (computer organization) evolved. But, it didn't evolve through the preservation of variations because of their consequences. Apart from that, the main thing on which I feel I differ from you is that sociobiology leaps a little too cavalierly from socio – to bio – . Sociobiology seems to leave me out. I'm in the middle.*

The "consequences" of which Skinner speaks are environmental imperatives that bring selection pressures to bear on organisms as they negotiate their world. Those physical attributes and instinctual behavior selected over enormous periods of time either facilitate survival or the organism meets its consequent. Additionally, Skinner guided us to understand his second kind of selection – the rapidly occurring selection process of operant behavior (Skinner, 1981). Organisms either operantly emit appropriate survival behavior (and live to see another day) or perish. Those behavior that result in survival are likely to recur; they must be shaped and conditioned quickly. Thus, *time* is the critical element frequently overlooked when (1) describing the nature of environmental consequences over considerable time driving evolution and (2) environmental consequences in very brief periods of real time driving operant conditioning and organisms' successful operant navigation of the world.

"Sociobiology seems to leave me out." This statement by Skinner is an important pronouncement from the premier radical behaviorist that sociobiology must consider the importance of the "role of variation and selection in the behavior of the *individual*," his second kind of selection by consequence" (Skinner, 1990). He suggests here that Wilson jumped directly from Darwinian natural selection to selection of social and cultural behavior, while skipping the intervening step of selection of individual behavior by consequences. Unfortunately, Skinner never directly articulated a complete conceptual framework that would provide biologists the appropriate bridge from their world of long-term Darwinian selection to his short-term and real time operant selection. He comes closest to achieving that bridge in both "Selection by Consequence" (1981) and "Can Psychology Be a Science of Mind?" (1990). Chapter 3 offers a much more comprehensive discussion of this concept.

6 . . . *Sociobiology seems to leave me out.*

When Skinner suggested that Wilson "leaves me out," he implied that biology failed to recognize the magnitude of his contributions with the development of the operant conditioning paradigm and its power as a biological agent. Skinner tried to draw a clear distinction between the much more mechanistic and time-dependent selection by consequences exerted on the physical manifestation of evolution (at the gene, individual, and species levels) and the more subtle and quickly developing "second kind of selection by consequences" of behavior and behavioral repertoires occurring within the individual across the lifespan, with the capacity to advance those repertoires to subsequent generations via culture – Skinner's third kind of selection by consequences.

Skinner's comments here anticipated Wilson's discussion of the selection of social behavior on the species level. That is, patterns of behavioral repertoires that advance the survivability of the group will be apparent to us as culture. The biological imperative to get our genes to the next generation imposes huge pressures on individuals to acquire behavioral patterns compatible with group norms (culture). Yet, in order to transmit culture from generation to generation, humankind had to possess a brain that included the capacity to develop the representational means (language and symbol) to codify (reify) cultural norms for others. That brain also had to possess developmental plasticity that enabled an emergent mechanism of behavioral adaptability to selection pressures that could occur within a single lifetime.

Skinner's operant conditioning is that mechanism, allowing behavior or behavioral patterns to be reinforced (selected), while not becoming maladaptive in the stereotypic stimulus-response fashion of classical conditioning (dealing with only "involuntary" behavior). Operant reinforcement and shaping of new behavioral patterns *do* change culturally transmitted repertoires and *do* interact with genes, as those genes which predispose an organism to behavior more likely to be reinforced are themselves more likely to be passed on to the next generation. Reinforcement of those behavior establishes a consistent pattern of tendencies in behavioral repertoires that serve to reinforce and select neural modules more capable of the unique processing demands of that behavior.

Please refer to Chapter 3 for a more comprehensive discussion of these concepts.

7 *. . . Do you ever stop to wonder about the intervening steps in the evolution of complex behavioral patterns? Evolutionary steps at every level that would have to be selected by consequences. Now isn't that a real problem?*

This assertion buttresses Skinner's 1981 *Science* article where he identifies three kinds of selection by consequences: natural selection, operant selection of individual behavior and operant selection via group contingencies of social/cultural behavior for the good of the species. He was building toward a discussion of how environmental consequences can operantly shape successive approximations of complex social behavior.

Skinner was apparently suggesting that such operant shaping is produced by environmental consequences that naturally select behavioral repertoires that enhance survivability. Organisms with a biological predisposition to be sensitive to the subtleties of environmental contingencies will more likely pass that tendency to offspring. He did not suggest that a particular stimulus will cause a determined social response. Instead, he seemed to have suggested a very malleable, yet powerful, biological capacity to be more efficiently shaped by environmental consequences. Such organisms will be capable of perceiving cues (discriminative stimuli) in their environments that will release the appropriate social behavior from among the learned repertoire. Discriminative stimuli, cuing and imitation are critical characteristics of the operant conditioning paradigm for the evolution of these social behavior repertoires. Each of the foregoing concepts is described from a Skinnerian perspective in Chapter 1 and again in Chapter 5 from a more biological perspective.

8 *. . . You are working on yourself to produce an act.*

Skinner's reference to "working on yourself," was his way of describing the existence of mechanisms that relate to humans preparing to emit a behavior. "Working on yourself," invokes a mental state by which humans rehearse the anticipated consequences of prospective behavior they might emit. Skinner asserts that the anticipation is based on individual reinforcement history related to a former environment that is now being generalized to a current experience. Presumably, internalized language (thought) is essential to this function (see his chapter: *Thinking* in his 1957 "Verbal Behavior").

9 *. . . I think you can always show that it would be to the advantage of the species if those stimuli which are now releasing behavior proved to be reinforcers.*

The stimulus that releases the initial involuntary behavior (S-R) becomes a potential primary reinforcer for operant conditioning of an emitted free operant behavior (in Skinner's ant example, the leaf that released the biting behavior may now serve as a primary reinforcer for operant conditioning). Skinner suggested that this evolutionary event, wherein even the simplest organisms developed responsiveness to the contingencies of operant conditioning, is of fundamental importance.

10 ... *We see X number of behavioral acts performed and we see the circumstances (discriminative stimuli or cues) under which they are performed. We note that only a very narrow range of circumstances evoke the behavior.*

Clearly, Wilson has accepted Skinner's suggestion regarding the importance of responsiveness to operant conditioning as an evolutionary event – Skinner's second kind of selection by consequences. If Skinner had taken liberty to insert the proper technical terms of operant conditioning, he might have corrected Wilson to say ... environmental cues (discriminative stimuli) set the stage for an organism to emit an appropriate behavior that may result in a primary reinforcer, thereby increasing the probability the organism will emit the behavior again.

11 ... *and that is searching behavior. Searching for what?*

Wilson was describing free operant behavior in his attempt to describe "searching." Reinforcement history will make particular behavior more likely as the organism operates in its environment. The range of those free operants (what Wilson seems to be referring to as "searching" behavior) will be further guided by the discriminative stimuli available in the environment. Organisms will engage in free operant behavior in the absence of stimuli and will certainly engage "searching" behavior in the face of the discriminative stimulus of appetite (hunger).

12 ... *there is a certain purpose element in the word "search" that I don't think belongs there.*

Skinner is correct to clarify purposeful (or search) behavior as different from free operant behavior. This basic principle is also consistent with Thorndike's Laws of Effect and Multiple Responses. Purpose would assume intent on the part of the organism. Skinner would argue that living organisms will do something rather than nothing – activity itself is reinforcing for living organisms and should not be mistaken for purposeful intent. Doing something simply means emitting free operant behavior. Free operant behavior is too often interpreted by most as purposeful, since we have a tendency to project human "thought" onto all other living organisms.

Starving, or the lack of appetitive satiation, can be quickly learned as a discriminative stimulus that makes more likely a larger range of free operant behavior being emitted among the ants – Thorndike's Law of Multiple Responses prevails as do Skinner's operant concepts of discriminative stimulus (cue) and free operant behavior. However, when hunger is introduced as a discriminative stimulus, it is likely to lead to free operant behavior that has achieved reinforcement in the past.

13 ... *tend to bring organisms to a relatively narrow set of accomplishments.*

The canalization of behavior to which Wilson refers is critically important for arguments regarding the nature of operant shaping of behavioral repertoires that become an essential component of the gene-culture coevolution – ultimately expressed through epigenesis and becoming increasingly represented in the observed species' social and cultural behavior.

14 ... *And then show that it does that (the ant moves toward the trail) statistically more than it would do if there were no trail there.*

This particular case is utilizing the light as a discriminative stimulus making it more likely the ant will emit the target behavior of moving toward the path, which, once established via reinforcement, will become an operantly conditioned behavior.

15 ... *There is no limit to what you can shape by way of complex behavior.*

Skinner determined that as long as an organism is physically capable of emitting the behavior, it can be operantly shaped to perform all manner of behavior it normally would have no purposeful use to perform. However, this basic understanding has significantly advanced the accomplishments of animal trainers who operantly shape animals to perform for the delight of many humans.

The case of salmon leaping is an example of the natural environment operantly shaping a behavior.

16 ... *You must have reinforcing consequences at every stage.*

Skinner was referring to the concept of successive approximations to the terminal behavior. In other words, a good behaviorist is capable of understanding the very basic elements of the terminal behavior desired. For example, a rat must first look toward a switch before it can be expected to press the switch to achieve a food pellet. Once the rat looks toward the switch (with reinforcement provided immediately), the reinforcement contingency can be increased to expect the rat to turn its body toward the switch, and then to move toward the switch, etc. Thus, the gradual increase of behavioral expectation successively approximates the terminal behavior until such time that the behavior occurs.

17 ... *It's too big a thing to occur at the level of (genetic) variation.*

These are not random variations of behavior somehow codified in the genome that result in a behavioral repertoire conducive to species survival. Skinner saw operant shaping as the mechanism to explain such behavior.

18 ... *I read Loeb in college and was lucky enough to get Crozier here.*

Historical note – Skinner was guided to read Loeb as an undergraduate student at Hamilton College. Crozier ran the biology laboratories in which Skinner found himself as a new graduate student. Crozier was much influenced by the work of Loeb and obviously reinforced that disposition in his new graduate student.

19 ... *There are gaps in my account too.*

Skinner made a significant and positive gesture to the work underway in biology and neuroscience to identify the source and vagaries of free operant behavior. It is also apparent that Skinner suggested that behavior never occurs in isolation; rather, it occurs within an individual having lived a unique array of experiences that shape a unique neural architecture, AND a social and/or cultural context ... "I don't know what happens in between." He recognizes that radical behaviorism was effective only in describing the overt characteristics of operant behavior and its external control, while neglecting the opportunity to suggest that the variables of daily experience shape observed individual differences in day-to-day behavior. Such individual differences are the stuff of behavioral neuroscience investigation that seeks to tap the neural processes that guide behavior.

Neuroscience researchers have significantly advanced our understanding of these covert brain/behavior mechanisms due in large part to applications of Skinner's operant paradigm in research protocols that link the measurement of performance on overt behavior with brain processing – via electrophysiological measures, such as event-related potential protocols or fMRI measures.

20 ... *when they became slightly reinforcing, that made it possible for new forms of behavior to appear in the life of the individual.*

Neuroscience researchers have significantly advanced our understanding of these covert brain/behavior mechanisms due in large part to applications of Skinner's operant paradigm in research protocols that link the measurement of performance on overt behavior with brain processing – via electrophysiological measures, such as event-related potential protocols or fMRI measures.

Skinner missed an opportunity to develop a stronger biologically based argument by neglecting the overarching relatively ultimate cause that he implied in his 1981 "Science" article regarding the universal responsiveness of all organisms capable of emitting behavior to be operantly shaped. Any organism incapable of responding to primary reinforcers (basic survival needs) will not survive to pass on genes to the next generation. Primary reinforcers have reinforcement value because they meet a basic requirement of survival. Organisms that emitted behavior resulting in increased access to primary reinforcement were operantly conditioned to emit those behavior again. In the strictest evolutionary sense, when emitted behavior increases access to resources in the environment (primary reinforcement), those organisms are most likely to live to see another day and might well succeed in passing their genes to the next generation.

Relatively ultimate cause in the case of this evolution of behavioral repertoire is the naturally selected susceptibility to be operantly conditioned possessed by all organisms capable of emitting behavior. The capacity to be operantly conditioned has an evolutionary history and has been incorporated into the system through many thousands of generations of natural selection. Relatively proximate cause in this primitive case is the behavioral repertoire (behavioral phenotype) that made it likely a free operant behavior might be emitted that would result in primary reinforcement.

Mayr (1961) introduced the concepts of ultimate and proximate cause in biology as a means to distinguish causes regulating responses to immediate factors in an environment (relatively proximate) from causes related to evolutionary selection that are represented in genotype (relatively ultimate). Further distinction regarding relatively ultimate and relatively proximate cause has been recently offered by Sober and Wilson, 1998 and Amundson, 2005. (See Prologue for complete development of that important distinction).

21 *...Then you get to the point where cultures prepare only for a world like that (which did the) selecting in the past. Then we are stuck with governments, religions, and capitalistic systems (that are more responsive to environments of our evolutionary past).*

These comments from Skinner are perhaps his most frequently rehearsed lesson to humankind. That is, we must understand our current behavioral repertoires are derived from our past reinforcement histories and not necessarily responsive to current environments, much less future environments. Pair that profound behavioral disadvantage with an ancient brain architecture evolved for responsiveness to the African savannah and we should achieve a better understanding of humanity's current survival predicament.

22 *...And so you've got three kinds of good there.*

Skinner's three kinds of good are: (1) what is good for the species promotes the genes, (2) what is good for the individual produces an effective behavioral repertoire, and (3) what is good for culture enables a group to survive. He more elegantly elucidates these in his seminal 1981 article, "Selection by Consequences," wherein he states: "human behavior is the joint product of (i) the contingencies of survival responsible for the natural selection of the species and (ii) the contingencies of reinforcement responsible for the repertoires acquired by its members, including (iii) the special contingencies maintained by an evolved social environment."

Chapter 5
To What Beginning?

> *A scientific view of man offers exciting possibilities. We have not yet seen what man can make of man.*
>
> *(B.F. Skinner, 1971)*

B.F. Skinner and E.O. Wilson both had long-term ambitions that their scientific contributions would be accessible to and understood by significant numbers of the public. It was readily apparent to each of them that they had a responsibility to appeal to a larger readership than the academic community. They also hoped that many readers might be motivated to make general application of their concepts to guide the emergence of a more altruistically responsive and peacefully sustainable world. Thus, a significant portion of their writing was given to more macroscopic issues regarding the relationship of their theories to the fate of humankind and to know human nature more fully for appropriate social planning.

Skinner and Wilson would likely agree that scientific explanations of human nature would serve humankind better than the unscientific approaches of ethics and some social sciences. Far from reductionistic, scientifically informed knowledge of human nature promises to advance the survival prospects for the human species. As David Barash (2005) has rightfully pointed out, a scientific conception of behavior "does not dehumanize man, it dehomunculizes him." Although far more typical among the scientific community today, this larger objective was uncommon during the earlier portions of their professional careers. For both, this aspiration occasionally served as the basis for criticism of their work and accusations that they were naïvely seeking to scientifically manipulate the social fabric in harmful ways. Indeed, Wilson writes in *The Naturalist* (1994), "At what point should scientists become activists? The ground between science and political engagement is treacherous. Speak too forcefully, other scientists regard you as an ideologue; too softly and you duck a moral responsibility." Fortunately, they both persisted in guiding the larger reading public to a more keen understanding of human nature. The results were books such as Skinner's *Beyond Freedom and Dignity* (1971), his utopian novel, *Walden Two* (1948), and various articles, such as "Evolution and Man's Progress" (1961) and "Can Psychology Be a Science of Mind? (1990)." Wilson produced works such as *Promethean Fire* (1983), the Pulitzer Prize winning *On Human Nature* (1978), *Biophilia* (1984), *Naturalist* (1994), *Consilience*

P. Naoutr, *E.O. Wilson and B.F. Skinner*, Developments in Primatology: Progress and Prospects, DOI 10.1007/978-0-387-89462-1_5,

(1998), and, his more recent, *The Creation* (2006), to name a few. In their own way, each of these works achieved an uncommon challenge to their readership – namely, challenge your assumptions regarding the place of humankind in the great chain of being and consider the means by which humankind can achieve a sustainable planet.

Skinner's Technology of Behavior

Skinner's unfailing aspiration as a radical behaviorist was for humankind to utilize the proven principles of operant conditioning to intentionally create "ideal" social structures that would maintain the most productive behavioral patterns. He was convinced that an empirical approach to a technology of behavior that provided a behaviorally designed culture would achieve that ideal society. He often wrote about his concern that individual behavior within society results from the most remote consequences to behavior – consequences that frequently extend far beyond the present generation. He identified overpopulation, pollution, and conspicuous consumption as examples. We all can easily acknowledge the longer-range consequences to these behavior. However, operant conditioning irrefutably demonstrated that consequences must be contiguous with an individual's behavior in order to achieve the desired change. In fact, the immediate gratification that comes from many of our worst behavioral practices serves only to maintain maladaptive behavioral patterns and further challenge the prospects of humankind's long-term survival as a species.

Skinner challenged us to consider designing social structures that craft "surrogate" positive consequences to provide meaningful contiguous reinforcers that would establish significant patterns of adaptive behavior that are incompatible with the former maladaptive behavior. Obviously, the utopian society he created in *Walden Two* was his effort to describe a society based on such surrogate positive consequences. In fact, the social structure of *Walden Two* is absent traditional aspects of the very institutions Skinner maintained provided the worst examples of meaningful behavioral structures – the negative reinforcers of government, religion, and education, and the contrived positive reinforcers that promote overconsumption in a capitalistic society.

Skinner effectively argued that our traditional social structures have made nearly exclusive application of aversive controls to manage behavior – punishment and negative reinforcement. Empirical data have unequivocally demonstrated that such consequences might achieve a short-term desired change of behavior, but the data also demonstrate that changes are short-lived. The other common result of such consequences is to establish a pattern of escape or avoidance behavior. Skinner makes a strong case in *Beyond Freedom and Dignity* that the Western concepts of free will and dignity are ineffective philosophical arguments in support of concepts that don't exist, concepts which are better explained by institutionally conditioned patterns of escape and avoidance.

Skinner was persistent about his technology of behavior. However, he was not so naïve as to think the social designs he proposed would be accomplished. Such designs must be provided . . .

> *by those who can do so effectively. That means governments, religions, and economic enterprises which control most of the conditions under which we live. They, however, are under the control of consequences controlling their own survival, which are much less remote and hence more powerful than the survival of the species. Moreover, the effects of these consequences are usually in conflict with it. For example, the legislator who sponsored a proposal to lower the birthrate, limit personal possessions, and weaken national and religious commitments would soon lose the power to sponsor anything. Business and industry cannot turn to the production of goods and services which will have fewer harmful consequences but will be less reinforcing to those who buy them. Religious leaders must make sure that their advice will be taken, and communicants will not take it if taking other advice has cost them reinforcers. Those leaders whose advice concerns consequences in another world must treat this world as expendable.*
>
> *The question is this: Under what conditions will those who have the power to control human behavior use it in ways that promise a better future?*
>
> *(Skinner, 1989)*

That power might reside in our species making a determined effort to understand the biological basis of our human nature, our relationship to the environmental contingencies that shape it, and deciding what to do about bringing biological and environmental variables under scientifically informed control. At the same time, "(i)t might prove to be that for a valid perspective of behavior to be incorporated into the fabric of world societies, it will have to be coupled with a social philosophy . . . that will provide/define 'purpose' for existence to the individual" (Rumbaugh, 1984). To his credit, Wilson approaches that question in a more significant way as he develops his concept of "consilience."

Wilson's Consilient Worldview

E.O. Wilson has been a synthesizer from the beginning. He stands among a pantheon of leading voices who continue the important tradition in evolutionary biology to advance beyond the mid-twentieth century modern synthesis, to the new synthesis of sociobiology. His stature was reinforced by his most recent and comprehensive consilient synthesis (1998). "To what end?" he asks in the final chapter of *Consilience*. In response, he suggests that the successful unification of the natural sciences, social sciences, and humanities will result in a revitalization of higher education in the liberal arts tradition. A tradition inspired by the so-called great questions – "What are we, Where do we come from, How shall we decide where to go?"

> *The future of the liberal arts lies, therefore, in addressing the fundamental questions of human existence head on, without embarrassment or fear, taking them from the top down in easily understood language, and progressively rearranging them into domains of inquiry that unite the best of science and the humanities at each level of organization in turn.*

Further, Wilson then encourages us to seek a synthetic disposition as we contemplate those great questions. Wisdom, he suggests, is achieved by people able to synthesize, "people able to put together the right information at the right time, think critically about it, and make important choices wisely." He continues,

> *The liberal arts will succeed to the extent that they are both solid in content and as coherent among themselves as the evidence allows. I find it hard to conceive of an adequate core curriculum in colleges and universities that avoids the cause-and-effect connections among the great branches of learning – not metaphor, not the usual second-order lucubrations on why scholars of different disciplines think this or that, but material cause and effect. There lies the high adventure for later generations, often mourned as no longer available. There lies great opportunity.*

As we apply Wilson's consilient worldview, his greater challenge to us is to consider appropriate responses to what he characterizes as two Mephistophelean bargains – each portending enormous implications for our future as a species. What might we *intend* to do in response to these bargains? Or, do we intend to continue our uninformed "shamble" toward our demise?

Our first choice results from an acceptance long ago of what Wilson identifies as the "Ratchet of Progress." The more we know, the more we can flourish, increase our numbers, and alter the environment – an outcome in itself that demands new knowledge just to survive.

> *In a human-dominated world, the natural environment steadily shrinks, offering correspondingly less and less per capita return in energy and resources. Advanced technology has become the ultimate prosthesis ... So to understand why humanity has come to relate to the environment in this way is more than a rhetorical question. Greed demands an explanation. The Ratchet should be constantly re-examined, and new choices considered.*

Wilson's second Mephistophelean bargain requiring our intentional consideration is directly related to the central concern of this book. The human species has achieved knowledge regarding what Skinner describes as three kinds of selection by consequences: Darwinian selection by natural selection, real-time operant selection of individual behavior, and selection of individual, social, and cultural behavior via group contingencies – the last, is intimately related to Wilson's gene-culture coevolution.

Darwinian selection has resulted in the evolution of life as we know it. Wilson asks,

> *Is natural selection still operating to drive (human) evolution? Have we in many ways exempted ourselves from the process of natural selection? Is it forcing our anatomy and behavior to change in some particular direction in response to survival and reproduction?*

He suggests a lot of "no" and, just perhaps, a small bit of "yes" in response to these questions. Over our history as a species, humanity has already likely responded to the natural forces that have challenged our survival, "overpopulation, famine, war, outbreaks of infectious diseases, and environmental pollution." Essentially, we have sat idly on the sideline and passively observed the results of Darwinian selection during the last century.

The very same disposition of passive observation has prevailed in the face of our increasing knowledge regarding the implications of Skinner's second and third kinds of selection by consequences. Do we have no intention to utilize the wisdom derived from the details of our knowledge to make informed choices regarding purposeful structuring of such selection mechanisms? Similarly, do we have no intention to utilize that same knowledge to make an informed choice to deliberately *not* impose such structures? Presently, our only obvious intention is to allow these processes to unfold absent the empirical wisdom we could bring to bear on prospective directions.

Certainly, Skinner offered many suggestions to his readers regarding intentional responses to the Mephistophelean challenge to utilize a technology of behavior to achieve more productive individual behavior and social structures. Obviously, we have mounted no significant response to his suggestions. Nor have we responded to the prospect that more beneficial social and cultural structures could result from intentional management of group contingencies described in Skinner's third kind of selection by consequences.

Wilson more bluntly articulates the essence of the second Mephistophelean challenge by rightly suggesting that we will soon have the capability to "alter not just the anatomy and intelligence of the species but also the emotions and creative drive that compose the very core of human nature. The prospect of this *volitional evolution* – a species deciding what to do about its own heredity – will present the most profound ethical choices humanity has ever faced."

> *Evolution, including genetic progress in human nature and human capacity, will be from now on increasingly the domain of science and technology tempered by ethics and political choice. We have reached the point down a long road of travail and self-deception. Soon we must look deep within ourselves and decide what we wish to become.*
>
> *Alter the emotions and epigenetic rules enough, and people might in some sense be "better," but they would no longer be human. Neutralize the elements of human nature in favor of pure rationality, and the result would be badly constructed, protein-based computers. Why should a species give up the defining core of its existence, built by millions of years of biological trial and error?*
>
> *What lifts this question above mere futurism is that it reveals so clearly our ignorance of the meaning of human existence in the first place. And illustrates how much more we need to know in order to decide the ultimate question: To what end, or ends, if any in particular, should the human genius direct itself?*
>
> *(Wilson, 1998)*

Chapter 6
The Challenge: A More Integrated Approach to Human Nature

It deserves notice that, as soon as the progenitors of man became social (and this probably occurred at a very early period), the principle of imitation, and reason, and experience would have increased, and much modified the intellectual powers in a way, of which we see only traces in the lower animals. Apes are much given to imitation . . . and that after a time no animal can be caught in the same place by the same sort of trap, shews that animals learn by experience, and imitate the caution of others.

(Darwin, 1871)

The 1987 conversation between Skinner and Wilson is a model of mutual intellectual regard and academic civility. It is one that embodies an ambition to achieve scientific progress via cumulative theoretical development. The conversation represents the best thinking of two widely acclaimed scholars as they sought to identify common ground in their theoretical positions at a particular moment in the history of science. Particular note should be made of the wide-ranging capacity of each to draw on their enormous wealth of scholarship and work toward a synthesis in discussing mechanisms of natural selection, operant selection, and cultural evolution. A synthesis was not achieved in the short time they had together, nor was it likely that one conversation could ever result in such an accomplishment – their conversation was never revisited. Thus, where does one go from here? What purpose is served by attending to their conversation at all? Might research findings since their conversation and more contemporary conceptual developments provide an even more consilient framework? How is their conversation relevant as important foundation for the next steps? "To what end, or ends, if any in particular, should the human genius direct itself?" (Wilson, 1998).

Readers are encouraged to view the conversation as one that provides a point of departure and goes far in providing the conceptual framework within which to consider a more completely integrated model of behavior – one that promises to be a useful end to which the human genius might direct itself (Rand and Ilardi, 2005). In order to be viable, such a model must be inclusive of Skinner's operants

P. Naour, *E.O. Wilson and B.F. Skinner*, Developments in Primatology: Progress and Prospects, DOI 10.1007/978-0-387-89462-1_6,

and hearken back to include Pavlovian respondents. It should also acknowledge the magnitude of influence culture (understood in Wilson's sociobiological framework) has in providing environments that produce rich experiences, functional behavioral cues, and effective reinforcements to behavior, while recognizing the significant bearing imitation has on initiating and enhancing learning. Additionally, the model should identify the developmental impact of early experience on the emergence of behavior. Finally, a more fully integrated model should offer a means to rise above reductionist tendencies typical of empirical practice and acknowledge that each of its elements is synchronously expressed as a function of naturally selected neurological mechanisms.

Duane Rumbaugh and his colleagues, David Washburn and William Hillix, offered a more integrated model of behavior ("Respondents, Operants, and Emergents: Toward an Integrated Perspective on Behavior") in 1996. The model, based on Rumbaugh's life work in comparative psychology and primatology, was cultivated over many years of meticulous behavioral research that earned worldwide recognition for his scientific contributions (reviewed in Washburn, 2007). The model warrants introduction in closing this text because it is so wholly based in behavioral psychology, while proposing a behavioral means by which the most sophisticated cognitive behavior emerge absent conditioning. Further, the model assumes a biological basis, and identifies the essential nature of the processes and products of culture that guide learning. Rumbaugh names these sophisticated cognitive behavior "emergents" and his integrated perspective as "rational behaviorism." The concepts are fully described in this chapter with the intention to inspire continuing and animated discussion among the wide range of disciplines focused on understanding behavior and "to achieve cumulative progress" (Rumbaugh et al., 1996). It remains for the individual practitioners within those disciplines to determine how their work supports this more comprehensive perspective. In Rumbaugh's (1996) own words, "A goal of good science is progress, whether reflected in cumulative theoretical development, or through Kuhn's (1962) paradigmatic revolutions."

This chapter begins by providing a brief review of the Skinner/Wilson conversation for context, continues by identifying key Skinner and Wilson concepts relevant to its purpose, and then offers overviews of the current research on (1) mirror neurons and imitation, and (2) developmental plasticity, evolution, and the emergence of new behavioral attributes. Those reviews are intended to provide additional relevant evidence to complete a contemporary context for offering a summary of Rumbaugh's integrated and more macroscopic perspective – rational behaviorism. His model is proposed as an appropriate linkage of respondents, operant selection by consequence, gene-culture coevolution, and emergents. Resolving that inclusive model holds the prospect of placing Skinner, Wilson, and Rumbaugh, and their theoretical work, in a much larger and well-deserved intellectual context. More importantly, an integrated reconciliation of radical behaviorism, sociobiology, and rational behaviorism has the potential to firmly establish evolutionary psychology as the common foundational element within the integrated behavioral sciences.

Skinner and Wilson

Skinner's apparent ambition from the outset of their conversation was to engage Wilson in serious consideration of the linkage between operant conditioning and mechanisms that lead to culture. He did so by first making absolutely certain that Wilson viewed as realistic the prospect that operant conditioning provides for two kinds of selection by consequences among the trio of selection mechanisms he had identified: (1) Darwin's natural selection, (2) contingencies responsible for selecting individual behavior and (3) special contingencies maintained by a social environment. In a much less overt fashion, he then attempted to sway Wilson to contemplate along with him how individual contingencies of operant conditioning can interact with group contingencies. Such an interaction can serve to reinforce individuals producing both interpersonal social behavior and engaging the behavioral repertoires of culture. Unfortunately, the conversation ended without Skinner resolving the relationship of operant selection with natural selection or suggesting their linkage to gene-culture coevolution. Identifying the (relatively proximate cause) mechanisms by which to close the biological loop among all three kinds of Skinner's (relatively ultimate cause) selection by consequence will operationally weave behaviorism into the theoretical fabric of gene-culture coevolution.

Enlarging the Conversation

Darwin's theory of evolution is driven by the long-term consequences of natural selection. Conversely, Skinner's real-time operant conditioning of new behavior is driven by the near-term contiguous environmental consequences to those new behaviors. Obviously, there are significant differences to recognize between the first two kinds of selection, in that Darwin's consequences are transmitted by reproducing survivors in the genes (genotype) and Skinner's consequences result in real-time changed behavioral repertoire (phenotype). Consequently, Skinner identified two significant limitations to his "second kind of selection by consequences." (1) The products of operant conditioning prepare an organism for a future that only resembles the recently selected past and (2) the size of an individual's behavioral repertoire is limited by the number of operantly conditioned behaviors possible for a single organism to acquire in a lifetime. However, he went on to suggest that "those limitations were corrected in turn by the evolution of processes through which organisms received *help* from other members of their species" (Skinner, 1989). The greatest portion of Skinner's "help" likely would be characterized by Wilson as the reified (symbolically represented) attributes of humankinds' eucultural (the most advanced form of cultural) existence – intentional modeling, verbal prompts, and teaching via direct verbal instructions – In other words, Skinner's third kind of selection by consequences.

The help to which Skinner referred is social behavior and the inception of culture. The initial help is that offered by a mediator who intentionally (or unintentionally) prompts (cues or models) a socially or culturally desirable behavior for another to imitate. Additionally, Skinner's statement suggested that a linkage exists between the operant conditioning of an individual's primed behavior, and both the mediator's operant modeling and reinforcement of such behavior. That linkage initiates an important social bond among humans, one that establishes the larger context for operant shaping of social behavior and, ultimately, gene-culture coevolution. The iterative relationship between the mediator and the imitator can be appropriately described as an operant feedback loop that engages both of Skinner's second and third kinds of selection by consequences. The process results in the acquisition of new behavior (social and otherwise) by one individual who effectively imitates the behavior modeled by another. Effective imitation is often reinforced by the person offering the modeling and frequently can achieve vicarious reinforcement within the person emitting the "new" behavior. Similarly, the mediator who provided the modeling achieves a reinforcing effect for the positive social behavior of "teaching" someone a new behavior. Such a process becomes increasingly rich in its interactive nature with the emergence of culture, the symbolic (reified) representation of prompting, and the more formally reified structures of schooling. Thus, the social dynamic of prompted imitation establishes the basis for closing the loop among Skinner's three kinds of selection by consequence.

The concept of imitation, both primed and self-initiated, has long been a topic of diverse representation in the research literature. It is among those concepts like "attention" that are presumed to have a common and well-accepted operational definition. Yet, scholarship suggests otherwise, oftentimes offering a simple and operationally useless characterization of imitation as a "combination of innate and learned behavior" (Rizzolatti, 2005). Rizzolatti goes on to suggest that Thorndike (1898) offered one of the more useful and parsimonious definitions: "learning to do an act from seeing it done." Thorndike's definition makes two assumptions, (1) imitation implies learning and (2) the imitator transforms an observed behavior into a volitional behavior that is identical to or closely resembles that observed behavior (Rizzolatti, 2005).

Thorndike did well by making the definition more precise, but his version offered little in the way of explicit operational features. Some years later, Skinner suggested a more straightforward behavioral approach to imitation that included operational characteristics and important mention of its evolved nature. His definition eliminated the notion that the individual was somehow "transforming an observed behavior" into a volitional act. Instead, he identified important operant features exclusive to human imitation uniquely produced by the imitator and the modeler. "Modeling is a way of showing another what to do. It primes behavior in the sense of evoking it for the first time and thus exposing it to potential contingencies of reinforcement" (1989). Thus, the imitator is provided contingent reinforcement by the modeler and the imitator's obvious intent to replicate the modeled behavior provides reinforcement for the teaching behavior of the modeler. Without question, human language emerged as an immensely powerful "tool" that provided a reified (symbolic/representational) means to describe behavior coincidentally with modeling, and to simply describe behavior without modeling.

The vocal behavior of telling very quickly became a much more effective means of priming and introduced a human interactive capacity that established an essential component of human culture, the teacher/learner relationship. Verbal prompting behavior quickly accelerated and was shaped by reinforcing consequences. Ultimately, we came to encounter much of our repertoire of social behavior via verbal injunctions such as proverbs, maxims, rules, and laws – all described by Wilson as reified elements of symbolically represented aspects of culture (euculture) that are unique to humans. Skinner goes on to say that

The origin and transmission of a cultural practice are thus plausibly explained as the joint product of natural selection and operant conditioning. A culture, however, is the set of practices characteristic of a group of people, and it is selected by a different kind of consequence, its contribution to the survival of the group.

Once imitation has evolved, contingencies prevail for the evolution of modeling. Operant behavior is imitated because the same reinforcing consequences are likely to follow. The imitation is important because it 'primes' behavior in the sense of bringing it out for the first time. Reinforcing consequences may then take over. (1989)

Skinner's (1990) willing inclusion of the evolutionary implications of operant imitation to human culture lends additional credibility to his contention that "sociobiology moves too quickly from the evolution of the species (his first kind of selection by consequences) to the evolution of cultures (his third kind of selection by consequences), by passing over a very necessary link between them, the operant behavior of the individual (his second kind of selection by consequences)."

The evolutionary implications of operant imitation identified by Skinner achieve their logical conclusion in his "black box." However, radical behaviorism largely neglected utilizing evolution within its conceptual framework and avoided addressing the biological basis for behavior. Consequently, radical behaviorism continued its existence in a silo, isolated from the rapidly emerging findings in cognitive neuropsychology and neuroscience. Ironically, those advancements were possible because researchers rigorously utilized discrete performance-based behavioral methodology to validate their hypothetical constructs. This same rigor has been (and continues to be) responsible for enormous advancement in our understanding of brain/behavior relationships over the past 25 years. Increasingly, most contemporary research on imitation has similarly positioned itself in a more inclusive context that intentionally explores its behavioral, developmental, cognitive, biological, and neurophysiological attributes. Hurley and Chater (2005) suggest that "recent work across a variety of sciences argues that imitation is a rare ability that is fundamentally linked to characteristically human forms of intelligence, in particular, language, culture, and the ability to understand other minds." This quotation precisely mirrors Rumbaugh's perspective in advancing a more integrated view of all behavior – without his inclusion of primate minds. Hurley's and Chater's two volume compilation of the work in imitation is a definitive collection of the major contemporary contributions in the field.

Some of the most exciting recent work in imitation has focused on the neuroanatomical mediators for imitation – mirror neurons for motor behavior and echo neurons for linguistic behavior. Obviously, neither Skinner nor Wilson had awareness of these neurons at the time of their conversation. However, this wide-ranging research now promises to go far in providing the basis for us to complete the loop that links Skinner's three kinds of selection by consequence. Additionally, these important biological mechanisms of imitation may provide the bridge that will enable linkage of Skinner's third kind of selection by consequence (group/social contingencies) to the evolution of culture and, ultimately, to rational behaviorism.

Mirror Neurons and Imitation

The discovery of mirror neurons in the early 1990s (reviewed by Rizzolatti and Craighero, 2004) provides the neurological link that promises to advance our understanding of the rich interaction between biology and environment, more precisely referred to as gene/culture coevolution. First identified in the monkey premotor cortex, mirror neurons appear to be the missing biological link to imitative behavior. The neurons have been shown to fire when a monkey *both* performs a specific object directed action *and* when the monkey observes another monkey or human perform that same action. In other words, a specific region of the monkey brain will show identical neural activity regardless of whether the action is being performed or observed. By logical extension, it is not difficult to imagine that the mirror neuron mechanism is very likely involved in some way with all imitative behavior. In fact, actual imitative behavior may be a neural inevitability – observed visual-motor performance information transformed into the observer's motor performance. Imitation is "released" by observation and processed within the interactive circuitry of mirror and motor neurons. Additionally, this mirror neuron mechanism likely influences an observer's capacity to actually "recognize" the actions of others and to verbally and visually represent the behavior after the *performance* fact.

Researchers have more recently focused on the prospect that the mirror neuron mechanism is an essential component in the development of even the most complex imitation/production interactions – human speech and language acquisition. The emergence of language related to the advancing development of mirror neuron function requires four features: (1) gesturing and meaning, (2) transfer of meaning gestures to orolaryngeal (articulation) gestures, (3) auditory mirror neurons, and (4) echo neurons (all reviewed in Rizzolatti and Craighero, 2004). Mirror neurons are the neural mechanism that links the sender of a gestural message with the receiver. Interestingly, the mechanism requires no immediate cognitive mediation, because its semantics are incorporated within the gestures used to communicate. Additionally, both word meaning and the motor mechanisms required for articulation are unrelated as we consider the emergence of speech. Thus, the evolution of speech requires the transfer of gestural meaning to abstract sound meaning – in other words, hand gestures and mouth gestures must be neurologically linked. Studies have verified that link and also shown it to include the orolaryngeal movements required for speech production. Combined with Skinner's long-identified concept regarding the operant control of the vocal musculature, the foundation is complete for human speech and emergent language structures and their neurological correlates.

The facial movements associated with eating provide the best example representing the transfer of gestural meaning to abstract sound meaning. Mouth, tongue, and lips all move in a specific fashion when we eat, and those movements constitute a gestural meaning known to everyone – eat. If, while producing those movements, we also generate sound, the characteristic, "mnyam- mnyam" or some variant like "myum-myum" is created. This near universal representation for eating shows how the meaning of a naturally understood action might be transferred to sound. It takes little imagination to suggest that auditory mirror neurons provided the next step toward speech acquisition when sounds originally accompanied by an action came to have the same meaning independent of the action.

Echo neurons provide the final component necessary to complete the transition from gestural meaning to articulated abstract meaning. The echo neuron system releases articulation when individuals listen to verbal material. Studies have demonstrated that humans stimulated with speech sounds will produce small muscle potentials in the speech articulation mechanisms – a kind of subvocalization. The system likely mediates imitation of speech sounds and, possibly, speech perception. Thus, the linkage of mirror neurons to echo neurons is complete. Mirror neurons provide the mechanism by which imitative motor behavior is released and meaning laden physical gestures are mirrored in articulation gestures that are then transferred to abstract sound meaning via auditory mirror neurons. Finally, echo neurons release articulation (or subvocalization) upon verbal stimulation of the auditory system.

This sophisticated neurological mirror system progressively releases motor behavior through the most complex echoic articulation behavior. The most meaningful theoretical implication of the mirror and echo system is the profound importance of imitative behavior within all organisms for the development of social behavior and, ultimately, culture. Humankind had the unique imitational attribute provided by echo neurons that interacted with the operant control of the vocal musculature to provide for language, a capacity that far outstrips the vocal signaling apparently the function of echo neurons in so many other species. This linguistic capability enables humans to "prime behavior by telling each other what to do as well as by showing them," and expanded cultural evolution beyond the range of any other species, by bringing "the human organism under a much more sweeping control of the environment" (Skinner, 1989). Certainly, the prospect exists that all mirror neuron and echo neuron related behavior are ultimately associated with social behavior and culture because they minimally require an imitating observer and a mediator who is performing a behavior. This dyadic relationship of the mediator and the imitator has been facilitated by the evolution of these special kinds of neurons, establishing the foundation for the interplay of gene-culture coevolution and the resulting emergence of epigenetic rules expressed in social behavior and culture.

The relatively proximate cause that provides the variation to drive the relatively ultimate cause of selection in this system is reinforcement of prompted or modeled behavior. In the most primitive evolutionary sense, organisms possessing neurons having a predisposition to function in the mirroring fashion were selected because they enhanced the likelihood for the occurrence of imitation. More importantly, the very nature of mirror neuron function suggests that imitative behavior functions analogously to a reflex arc, rather than being encumbered by layers of redundant neuronal circuitry and processing that require higher order learning. The deeper cognitive processing and linguistic association necessarily occur after the reflexive mirror neuron mediated imitation, much like the more meaningful perception of "that was hot, I won't do that again" follows the reflexive motor response and "ouch." This naturally selected mirror neuron-based imitation provides the biological bridge between operant reinforcement of those behavior within individuals and final expression in behavioral repertoires, social behavior and culture – Skinner's second kind of selection by consequence (individual contingencies) and his third kind of selection by consequences (group contingencies).

The biological bridge effectively closes the interrelated loop among Skinner's three kinds of selection by consequences. Moreover, these relatively proximate cause mechanisms provide the variation in behavioral repertoire upon which the selection process acts in gene-culture coevolution. Subsequently, the next logical consideration is the neural manifestation of this gene-culture coevolution that is required to organize the higher order cognitive processing and produce behavior. This neural development and organization is the product of the interplay among genotype, developmental plasticity, and neural selection in the face of individual and group environmental contingencies. Thus, expressed behavioral phenotype is not solely dependent on genotype – neither do genotype nor environmental stimuli cause automaton-like or stereotypic behavior. Instead, behavioral phenotype is linked to genotype by *development*. In reality, the evolution of development (in this case, neural development) may be synonymous with the evolution of phenotype (West-Eberhard, 2003).

Developmental Plasticity and Evolution

Until very recently, the manifest conceptual importance of development to evolution was, for the most part, shelved in deference to the new synthesis and its historical emphasis on genetics. Ignoring ontogeny was important for allowing geneticists to avoid Lamarkian errors, but it became so entrenched that developmental phenotypic plasticity continues to be among topics "most estranged from modern evolutionary theory and genetics" (West-Eberhard, 2003). However, the prospect is now becoming more apparent that our capacity to describe the ontogeny of neural mechanisms for gene-culture coevolution will likely be derived from advances in research regarding developmental plasticity and its potential relationship to evolutionary developmental biology (evo devo). The interplay of the two areas holds the possibility that new insights in the neural mechanisms of gene-culture coevolution might be based in embryology and ontogeny.

The foundation of evo devo is based on the premise that "neither natural selection nor DNA directly explains *how* individual *forms* (phenotype) are made or how they are evolved (Carroll, 2005)." The work is advancing rapidly in describing the developmental process by which a single-celled egg emerges over time to become a complete organism; however, current research has not extended to a comprehensive cataloging of human *form*. Is it reasonable to suggest that "tool kit" genes and genetic switches combined at a point in human evolution to provide for the extended neotenous *form* so critical to the truly emergent nature of human ontogeny? If so, is it also reasonable to suggest that humans' extraordinarily extended neural ontogeny enables sufficient time for the interplay of variables that result in human social and cultural behavior?

These questions are important to determine the relationship of evo devo to work in human developmental plasticity and evolution. The present gap in most evo devo work is its focus on "gross morphological variation across phyla, with little or no discussion of behavior, physiology, life histories, and the kind of variation within populations that is required for natural selection to work" (West-Eberhard, 2003). Such a significant focus on the genome leaves out essential considerations regarding environmental influence. This linkage to environment requires the perspective that development always begins with an inherited phenotypic course. West-Eberhard goes on to suggest that phenotype will be "highly responsive to new inputs, which can be either genomic or environmental. These inputs cause the responses that correspond to new switch or decision points in development and give rise to new phenotypic traits. The result is development of a phenotype that is a mosaic of switch-determined modular traits."

The remarkably immature state of the human brain at birth establishes a clear and powerful relationship with environmental consequences that play out over decades. Thus, behavioral phenotype evolution can be more clearly understood by explaining the evolution of phenotype development (gene-culture coevolution). In other words, the extended neotenous state of the human brain enables behavioral phenotype to emerge over decades of selection via individual and group environmental contingencies – the adaptive flexibility of the human brain is an expression of the evolution of phenotype development. Those critics of gene-culture coevolution who claimed it to be deterministic neglected the conceptual importance of the evolution of phenotype development – they too quickly skipped (most often simply neglected) development, routinely skipping from phenotype directly to genes (West-Eberhard, 2003). This problem is a likely consequent of the, as yet, unresolved epistemological problem that emerges from the traditional position of adaptionists, based in population genetics, and the structuralist position held by developmental biologists. Although beyond the scope of this book, an epistemological reconciliation of those two positions will go a long way toward placing the psychology of human nature within the framework of a fully synthesized evolutionary thought. In other words, we will reconcile the possibility that both populations (adaptionist) and ontogenies (structuralist) evolve, while contributing in important ways to phenotypic expression (Amundson, 2005).

The ontogeny of individual and, ultimately, species neural substrates that guide the expression of behavioral phenotypes, such as social and cultural behavior, is dependent on the interactions among genotype, and both individual and group environmental contingencies – otherwise described by E.O. Wilson as gene-culture coevolution. As well, the truly emergent nature of human ontogeny is further enriched by humankind's extended neotenous state. Together, these variables provide the requisite trajectory and flexibility for gene-culture coevolution and its expression in human social behavior and culture.

> *An adult's response to a stimulus is clearly a matter of proximate causation. However, the adult's behavior is only partly determined by the ultimate evolutionary origin of its genotype. It is also determined in part by the environmental influences on the adult during early development. The influences from early ontogeny are clearly "more ultimate" than its present stimulus, but not so ultimate as the evolutionary selection for its genotype. Why not conceive of ontogeny as an in-between point in the proximate - ultimate scale? (Amundson, p. 204)*

The fact that organisms (phylogenically) are susceptible to environmental contingencies of operant conditioning is a relatively "more" ultimate cause. The various behavioral outcomes of the operant conditioning are relatively "more" proximate as mechanisms of gene-culture coevolution. However, those relatively more proximate behavioral repertoires are essential proximate attributes upon which coevolutionary selection acts to provide those behavioral repertoires; thus, completing the cycle of interaction in gene-culture coevolution. The picture that has been created in this evolutionary scenario suggests that environmental contingencies control, both in a near-term and long-term fashion, the outcome we recognize as the behavioral repertoire called human nature.

Yet, Amundson's (2005) challenge to conceive of ontogeny "as an in-between point in the proximate-ultimate scale," leaves a compelling gap – perhaps, a real void – in a successful attempt to overcome the arguably mechanistic natures of radical behaviorism and sociobiology. A truly integrated reconciliation of behaviorism and biology must transcend the inherent reductionism in both paradigms. This chapter has identified some intriguing new empirical developments as a means to challenge readers to traverse discipline-based boundaries and consider relationships that might provide a more integrated and comprehensive view of behavior. Together, with the book's comprehensive attempt to reconcile radical behaviorism and sociobiology, these additional empirically based conceptual elements suggest that a case can be made for considering a more "rational and less radical" all-inclusive behavioral perspective (Rumbaugh, 2008a). This case must offer a behavioral mechanism for the many sophisticated cultural and cognitive behavioral repertoires that are described by gene-culture coevolution – most particularly, behavior that occur among primates and humans. The rational behaviorism and emergents of Duane Rumbaugh equip us with the necessary conceptual elements to suggest how the processes and products of culture combine with the neurobiology of the brain to enable complex organisms to produce cognitive and cultural behavior – behavior that are not conditioned in the traditional behavioral sense.

Rational Behaviorism and Emergents

Rumbaugh's notion of rational behaviorism had its foundation in the work of Harry Harlow. Indeed, Rumbaugh credits Harlow with "breaking the bonds of Thorndike's psychology that strictly bonded responses to stimuli as a consequence of rewards" (Rumbaugh, 1997). More to the point, Harlow broke those bonds and initiated the emergence of rational behaviorism when he argued that the marked improvement in his monkeys' learning skills in the laboratory constituted a kind of insight, or "Ah ha!" capability. His observations inspired him to subsequently acquire a group of naïve monkeys who had no prior, and potentially biasing, laboratory learning experiences and to engage them in a full experimental protocol to test his earlier observations. Harlow's data were published in his groundbreaking *learning sets* article (1949). The results, as summarized by Rumbaugh and his colleagues (1997), indicated that

> *... despite limited, indeed insufficient opportunity to learn each of a long series problems, there emerged lawfully, gradually, and predictably, a new style of learning by Harlow's monkeys. Their learning shifted from gradual intra-problem improvement on the first few discrimination problems to insightfulness, where each new problem was learned to near perfection in one trial – the first trial. The rhesus had become transformed from a Thorndikean trial-and-error learner into a one-trial, insightful problem solver.*

When animals achieve a cumulative experience with a significant number of problems comprised of a single class, Harlow suggested that a more Gestalt learning model prevailed and subsequent novel problems within that same class would be solved insightfully and "supplanted by an orientation toward task accomplishment via creative actions, afforded by emergent processes that generate new options of choice both behaviorally and intellectually" (Rumbaugh, 1997). In a very real sense, Rumbaugh viewed Harlow's observations as an indication of emergent *thought*.

Harlow's "learning set" concept animated Rumbaugh's thinking for decades as he engaged a hugely productive primate research program. Finally, he and his colleagues were inspired (Rumbaugh et al., 1996) to suggest that prevailing twentieth century behavioral concepts of respondents and operants didn't explain such *thought*, leading them to propose an entirely new class of behavior – emergents. This third class of behavior provides for a "trichotomous classification of behavior that recognizes and adds to Skinner's (1938) distinction between respondent and operant conditioning" (Rumbaugh et al., 1996). Additionally, emergents serve as the foundation for his more integrative behavioral model – rational behaviorism, which "allows for the emergence of new behavior generated by cognitive operations of the primate brain ... reflect(ing) the natural operations of the brain as being comprised of keen pattern-detection and synthesizing systems" (Rumbaugh, 2002).

Traditional behavioral concepts of respondents and operants serve to account for behavior noted for their predictability and fixedness. Each is patently well documented, has survived decades of thorough empirical examination, and can be observed across the entire spectrum of organisms capable of producing behavior. However, not all behavior can be so readily characterized within these traditional behavioral categories. Conversely, emergents are *not* noted for their predictability and fixedness. Rather, they are identified for their innovativeness and cleverness. Many species are capable of emergents, but the probability of emergents being generated is significantly related to the biological attributes of cerebral complexity and encephalization, characteristics that provide for requisite integrative and generative functions.

Although emergents have no identifiable history of specific training, such as the conditioning which characterizes respondent and operant behavior, they do draw upon, but are not the specific consequence to, the learning experiences that result from such conditioning. Accordingly, emergent behavior only partially reflect an organism's reinforcement history and incorporate conditioning as only an element of the whole.

Indeed, they are not the result of reinforcement in any traditional sense of that behavioral concept. Rather, emergents manifest the species′ adaptation to its ecological niche, and both its pattern-detection and synthesizing functions that serve its adaptation. Emergents draw upon sensory perceptions and general life experiences from birth onward. Therefore, an essential biological implication to note in this conceptualization is that emergents are configured by an organism's neurological processes; emergents do not represent the specifics of conditioning or reinforcement histories.

Thus, emergents include all patterns established in the organism's classical and operant conditioning histories in addition to all "experiences" of daily life. Rumbaugh and his colleagues propose that these wider ranging "experiences," along with those learned through the processes of conditioning, be identified as *amalgams*, which form in real-time on the basis of reliable event/stimulus contiguity. "(A)malgams spontaneously arise from one's perceptual experience with contiguous stimuli without necessarily requiring a reinforcement or unconditioned stimulus to complete the amalgamation" (Rumbaugh et al., 2008). The constituent components of the amalgams share their qualities and their response-eliciting properties according to their respective strengths. This model (Rumbaugh et al., 2007) supplants any need to consider special reinforcement effects other than to the degree that all reinforcers are strong stimuli, relative to other stimuli with which they co-occur, and, hence, have dominant influence as they enter into amalgam formation (Rumbaugh, 2008b). But how the brain of any given species produces emergents is not known. Rumbaugh posits that it is via the brain's continuous processing of amalgams that emergents are produced in order to resolve the dissonance among them – what we verbally refer to as logical systems. In other words, when proven behavioral systems are inefficient or no longer working well, what logical alternatives does an organism pursue in the interest of adaptation? (Rumbaugh, 2008b).

Emergents are defined as "new behavior patterns that are noted for being synergistic, integrative and clever." They manifest themselves "as new capabilities, such as speech comprehension, that are not to be accounted for as responses or behavior altered by basic conditioning procedures" (Rumbaugh, 2002). While emergents, by their very nature, include the entire array of an organism's behavioral and experiential history, Rumbaugh offers the following characteristics to further differentiate them from respondents and operants:

- *Emergents' initial appearances come as unanticipated "surprises" to the researcher.*
- *Emergents provide novel response patterns and solutions to problems.*
- *Emergents form covertly, hence unobtrusively or silently.*
- *Emergents afford new behavior that have no specific reinforcement history.*
- *The formation of emergents generally cannot be charted.*
- *The formation of emergents emphasizes classes of experiences.*
- *Emergents entail the synthesis of individually acquired responses and experiences.*
- *Emergents are not subject to specific stimulus control as are respondents and operants.*
- *Emergents frequently reflect rearing conditions and/or early experience.*
- *Emergents tend to be associated with brain complexity (as per species maturation).*
- *Emergents enable an organism to learn to use symbols as representations of things and events not necessarily present.*
- *Emergents enable comprehension and use of language, to speak and sing.*
- *Emergents enable learning vicariously from secondary records (written materials and other records).*
- *Emergents enable reflection upon past experiences and events projected in the future.*

(Rumbaugh, 2002 and Rumbaugh et al., 1996)

A notable aspect of Rumbaugh's conceptualization of emergents is his intentional inclusion of discussion regarding neurobiological properties that allow emergents to arise. Reminiscent of Donald Hebb's (1949) cell assembly concept, Rumbaugh and his colleagues (2007) offered a description of the neurobiology of sensory perception and synaptic activity that result in activation of the neural circuitry of the brain. Very simplistically, as learning events occur and cell assemblies strengthen their connections with related cell assemblies, learning can become a larger amalgam and become the basis for cognition. "The combination of cell assemblies, or the combination of associations between events, could then produce a higher cognitive function such as a concept, thought, or emergent behavior" (Rumbaugh et al., 2007).

Behaviorally, these neurobiological alterations can be identified as emergent response modes, while those who view them from a cognitive perspective might describe them as cognitive operations (Rumbaugh et al., 1997). Regardless, emergents always reflect extant biological properties; thus, might it be that their phylogenic appearance recapitulates the ontogeny of their human appearance? Rumbaugh and colleagues (1996) respond affirmatively to the question and further suggest that the full range of emergent behavior are only available to adult humans and are dependent on the timing and interaction of experience with biology (Wilson's epigenesis). They continue by adding that some organisms might never produce emergent behavior beyond Piaget's sensorimotor stage and that the higher Piagetian stages may arise in the behavioral repertoire of more complex organisms. Indeed, Rumbaugh's primate research demonstrates that some aspects of Piaget's formal operations occur in chimpanzees, orangutans, and gorillas who have achieved aspects of linguistic behavior. This emergent behavioral progression that defines Darwinian evolution provides investigators a comparative behavioral framework and serves as the basis for rational behaviorism. "the term *rational* refers to the inclusion of cognition; the term *behaviorism* acknowledges the fact that the only data available to our science are behavior" (Rumbaugh, 2002).

Rational behaviorism posits that the organism will, given the opportunity, attend to the consequences of its behavior. It will monitor the environment, its own behavior, and the behavior of others, and will be sensitive to the consequences thereof. The organism will behave and monitor its behavior to the physical and social environs as though it is alert to the detection of possible cause-effect relationships . . . (T)hrough their behavior, organisms strive for control over their environments and themselves. . . Control can be achieved in measure by what we will call biological smartness, but strategies and tactics in pursuit of control are perhaps most facilely designed and carried out by psychological intelligence. (Rumbaugh, 2002).

Final Thoughts

Rumbaugh's integrative wisdom completes the circle of interaction that begins with Pavlov's respondents, gains momentum in Skinner's operants, is reinforced by Wilson's sociobiology, and achieves conclusion in his own concept of emergents. He is correct to prominently acknowledge the primacy of experience, imitation, and learning to emergents; and, in the tradition of Harlow, he is wise to identify the essential nature of learning, neurophysiology, and encephalization (1996). This book has simply provided closure to his circle by offering an explication of the operant basis for the sociobiological evolution of culture, which products (culturgens) and processes (epigenesis) provide the rich interaction of experiences and biology that drive the formation of emergents.

Behavioral science will be well served in the years to come by utilizing Rumbaugh's more inclusive and integrated behavioral perspective – a perspective that views higher order organisms as rational agents, thereby supplanting the radical behavioral model of the twentieth century with a model that is based on complex learning, complex behavior, and rational "thought." The rational behavioral perspective is founded in the finest tradition of twentieth century behaviorism; it is inclusive of more contemporary work in cognitive psychology, sociobiology, evolutionary psychology, and neuroscience; and it offers behavioral scientists from all disciplines an all-inclusive comparative behavioral framework.

Epilogue: Some Historical Notes and Indications of the Future

(H. Carl Haywood, Vanderbilt University)

The Position of Darwinism in Contemporary Psychology

A college president once asked me whether I could teach psychology without mentioning sex or evolution. My reply was that I could but I would not. To do so would be scientifically and academically irresponsible, requiring me to ignore both a vital topic in human (and other animals') behavior and the single most productive conceptual underpinning of human development.

The science of psychology has many roots whose relative influence depends on who tells the story. Differential psychologists point to Galton and Binet as well as Maskelyne and Bessel. Psychotherapists inevitably mention Freud. Experimental psychologists continue to dispute priority: Wundt or James. In my own mind, by far the most important event in the history of psychology was Darwin's (1859) publication of *On the Origin of Species*. Pavlov's demonstrations of conditioning of reflexive behavior in dogs would have had far less meaning, and no more than an analogic application to human behavior, had it not been for the scientific community's near-unanimous acceptance by that time of Darwin's theory of evolution by natural selection. Ernst Haeckel (e.g., 1894) could not have made his famous dictum on development, "Ontogeny recapitulates phylogeny," in the absence of an evolutionary base for developmental sciences, and our understanding of individual development would have been the poorer. Even though that assertion is not strictly true, it brings force to the important idea that neither evolution nor ontogeny is random, and that they are conceptually related in ways that enrich the understanding of both. The entire enterprise of comparative psychology rests on an assumption of some measure of continuity in various life forms. If, in its broadest sense, the ultimate goal of psychology is to understand the human condition, then it is essential that we have some grasp on the questions of where we came from and how we got here. Darwinism has given us an enormous head start on answering those questions, and could even help us to divine where we are headed and how likely we are to get there.

Evolution by natural selection has become, in the century and a half since Darwin's original publication, the foundation and cornerstone of biological science. The present question is whether it is a similarly useful set of concepts for the development of psychology. Naour's (this volume) proposal of a confluence of radical behaviorism and sociobiology suggests not only that it is, but also that radical behaviorism shares with sociobiology a debt and an allegiance to Darwinism. The pressing questions for me then become the following: (a) In what sense does contingent reinforcement represent a survival mechanism? (b) What has been—and is—the role of reinforcement in human development and the "higher mental processes" such as learning, systematic logical thinking, and creative/innovative thinking? (c) To what extent, and with what profit, can the evolution of species and the development of cultures be seen as sharing basic processes and concepts? (d) How do we get all the way from conditioned reflexes to complex and innovative thought?

Problems with the Concept of Reinforcement

The concept of reinforcement is at the heart of these questions: its definition(s), its applications in research on learning and behavior change, and its future as an explanatory tool. The principal difficulty derives from the classical definition of reinforcement as reduction of primary or secondary drives, based in turn on biological needs. The biological problem with that concept is simply that quiescence, a "steady state," turns out not to be the compelling desired outcome of either behavioral or biological systems. The ultimate goal of behavior is not nonbehavior. A further problem is that concept's implicit denial of nonreactive behavior, that is, behavior whose predisposing and precipitating causes lie within the neural structure and experiential history of the organism.

The breakthrough physiological and philosophical work was Granit's (1947, 1977) discovery that retinal neurons do indeed fire in the absence of photic stimulation. The brain is an active organ. There were further demonstrations leading to the conclusion that the brain is programmed to be active rather than inactive. It is not difficult to imagine the survival value of an active rather than a passive brain! Other researches (e.g., on stimulus deprivation) led to the notion that when deprived of its individual optimal level of external stimulation, the brain will generate stimulation whose origin is inside the body, as shown by hallucinatory states in stimulus-deprived persons as well as such persons' swift intolerance of stimulus deprivation. Granit (1977) observed the grand philosophical nature of these events in his distinction between the older physiological concept of *adaptation* and the newer one of *adaptability*. The latter concept implies a dynamic neural/physiological system with constant readiness for change and, in fact, responding primarily to change. Such a dynamic system plays such an important evolutionary role that it can be generalized, at least by analogy, to social and cultural adaptability, including even geographic differences in psychological characteristics (see e.g., Rentfrow et al., 2008). Perhaps it is an acquisition of the *ability to adapt* to changing circumstances rather than adaptation itself that propels evolution of both species and cultures. If adaptation were the ultimate goal of development, we might all be opossums: animals that, until the advent of the automobile, were very well adapted to their environment, so much so that the species had not evolved in millions of years. Even so, it is difficult to see wherein contingent reinforcement propels cultural evolution. It is likely that neither physiology nor cognition is sufficient to explain such phenomena and that more powerful concepts of motivation, including motivational theories of cognition and cognitive theories of motivation, must be added to the philosophical mix. That possibility is explored later in this epilogue.

The active-brain breakthrough called into serious question the theories that rested on steady state assumptions, including reinforcement as drive reduction. It also laid the groundwork philosophically for explorations in cognitive psychology by suggesting that events can *originate* within the nervous system. Cognitive psychology, which had suffered neglect and disdain during the heyday of behaviorism, began to enjoy a renaissance.

Rational Behaviorism as the Next Step

Naour has been especially perceptive and forward-looking in placing rational behaviorism in the logical and temporal sequence of thought about intelligent behavior, after radical behaviorism and sociobiology. As much as the first two movements, radical behaviorism and sociobiology, seemed at first to be curious bedfellows, rational behaviorism might seem also to fit uncomfortably with both. It is their fundamental reliance on evolutionary concepts that provides the common basis for the three movements, but there is more than that. There is a progression from simple to complex explanatory principles, and ultimately from attempts to explain simple and single acts to the explanation of complex, creative, and original thought—with an apparent detour toward explaining the behavior and evolution not merely of individuals or even species but of cultures as well. This sequencing is at best an ordinal scale, more likely a nominal one, there being no chance at all of its representing equal logical intervals. Nevertheless, it is this kind of sequencing that can provide the springboard to the next generation(s) of conceptualizing and empirical research on intelligent behavior.

As is always true in the cumulative scientific enterprise, rational behaviorism would not have been possible had it not been for some prior developments in thinking about intelligent behavior. It is also the usual case in science that there were barriers to be thought through and overcome. One of rational behaviorism's virtues is the fact that it has not ignored or discarded its history, especially its debts to both classical and operant conditioning. There was overcoming to be done, especially with respect to the role of reinforcement. Some attention to its antecedents can help one to understand both the structure and the potential of rational behaviorism.

Sneaking Up on Cognitivism

It is a popular saying among graduate students that Sigmund Freud (and other determinists) took away the will from psychology and John B. Watson took away the mind. To the extent that it is true, what was left to study was just mindless behavior – of which one sees far too many examples every day! Fortunately, psychology never completely lost its mind. There has always been some explanatory reliance on "mental" processes. The complexity of human behavior could never be understood entirely on the basis of observable and directly measureable events. Learning, for example, has had to be inferred from observable changes in behavior, presumably as a consequence of specifiable experiences. Intelligence, to an even greater extent, is an example of such a latent variable: unobservable, unmeasureable by any direct means, inferred on the basis of achievement or performance on tasks thought to depend on "intelligence." We assess individual differences in such performances quite reliably so long as we stick to the manifest variables, but we still have to infer the existence of the latent variable. We *know*, radical behaviorists to the contrary notwithstanding, that events are taking place between identifiable stimuli, between stimuli and responses. My own acquaintance with B. F. Skinner convinces me that he understood this quite well but considered the issue a tactical one. Against a background of introspection, instincts, and mentalism, he believed that psychology's most profitable exercise would be to concentrate attention on directly observable events while meticulously specifying the conditions (in his case, schedules of reinforcement) whose engineering could alter behavior. His behavioral engineering has enjoyed considerable success, although I see it as an engineering tool rather than a scientific one in the sense that its success has been more in changing behavior than in explaining it. As an antecedent of rational behaviorism, Skinnerian behaviorism provided a means of eliciting different forms of behavior that might not be observable under natural conditions – such as getting chimpanzees and bonobos to manipulate keyboards marked with linguistic symbols in order to get things they want. Hull–Spence behaviorism, relying on the fundamental concept of contingent reinforcement, provided a further platform, but it was still a means of establishing habits and did not consider whether cognitive events might be requisite to the formation of those habits or possibly develop or emerge from them.

The classical Gestalt psychologists/philosophers, Wertheimer, Koffka, and Köhler (see, e.g., Koffka, 1935) had never been drawn into behaviorism; consequently, they were free to explore more complex behavior. Köhler's demonstrations of "insight" problem solving in chimpanzees, especially his dramatic demonstrations with the chimpanzee Sultan, provided evidence that new behavior could emerge as a joint product of prior experience, contemporary task demands, and a novel organization of cognitive processes. Thirty years later, Harlow (1949) showed that monkeys can build upon cumulative experience both to provide motives for work and learning and to yield innovative solutions to problems, as well as to establish "learning sets" or an apparent learning-to-learn phenomenon.

Finally, cognitive and motivational aspects of behavior change were coming together. Psychologists needed to identify *motives* for behaving in nonreactive (i.e., instigational) ways. At the same time, there was a clear need – well, not so clear at the time – to get beyond associative learning, or even learning itself, to explain behavior change, especially the leap from drive-reduction-dependent conditioned responses to complex chains of behavior and ultimately to new behavior.

Edward C. Tolman (e.g., 1932, 1958) was a very important figure in that journey. Rather than being content to study molecular behavior and reflexology, Tolman, influenced by the Gestalt psychologists and by McDougall, introduced "purposive behaviorism" in the study of molar behavior: the whole organism acting as a unit. He included instincts along with physiological need states in his concept of "demands." His most compelling and encompassing concept was "goal-directed behavior," according to which he claimed that one could observe behavior and see in it an organism's purpose or goal – that is, where the animal wanted to get to. Expectation was "the intentional forward-looking determinant which makes behavior goal-directed" (Boring, 1950). This well-respected psychologist was recognizing that animals (including human beings) behave with identifiable purpose, that purposive behavior can be identified with the aid of rewards that are important to the animals, and that goals can be anticipated not so much in behavior as in the cognitive – yes, cognitive – processes of the behavers.

Twentieth Century Cognitivists

During this time, a Russian psychologist, L. S. Vygotsky, and a Swiss epistemologist, Jean Piaget, were building conceptual systems about the development of systematic thinking. According to Vygotsky's social-historical perspective (Vygotsky, 1978, 1986), cultures contain all of the tools that children will need in order to achieve systematic, effective, and abstract thinking and to learn a wide variety of content. The developmental task of parents and other teachers is to acquaint the children with the preexisting cultural tools and provide opportunities for their mastery and application in problem-solving situations that can be novel. Thus, culture and cognitive development become intertwined. Karpov (e.g., 2005; Karpov & Haywood, 1998) has provided evidence of the success and efficiency of teaching cultural tools to children of different ages. If indeed cultures contain the tools that children need for systematic thinking and learning, one has to ask how those tools become embedded in the culture. Given that cultures depend to one extent or another on geographic, climatic, population, and many other environmental conditions, it is perhaps the adaptability principle that permits and drives cultural evolution. That is to say, the ability to adapt to different environmental demands – and benefits – is itself the survival mechanism that guides evolution of the species along different paths toward cultures. It is still a long stretch from the adaptability of individuals to the evolution of cultures, and it remains difficult to see the role of reinforcement in this process.

Piaget's (1950, 1952) developmental epistemology is the final forerunner, and contemporary link, which I wish to make to rational behaviorism. Contrary to Vygotsky, Piaget did not assume that the cognitive tools are present in the culture; rather, he assumed that each individual has the ontogenic task of constructing his/her own cognitive "structures," that is, sets of systematic, generalizable, and relatively permanent thinking modes. That constructive process occurs in an orderly and predictable sequence, with progressively more complex and abstract thinking modes developing by periods and stages, as a function of children's consecutive interactions with their physical and social environments and of a maturing nervous system. The essential elements of this approach, in the present context, are (a) the relation of knowledge (experience) to understanding and abstract cognition, (b) the manner of progression from stage to stage in development, and (c) the role of disequilibrium.

Piaget's fundamental notion that knowledge precedes understanding has profound implications. It gives primary importance to the role of cumulative experience, that is, knowledge as information about one's world (*des connaissances*). He insisted that the ability to comprehend the meaning of information (*la connaissance*, understanding) is, in very large part, a product of the accumulation of such knowledge. The steady acquisition of knowledge can advance cognitive development through the successive steps within a stage of development, and this progression can enable one to become more and more proficient at doing the same thing, that is, at carrying out a practiced act of thinking. Something more must occur in order to gain the ability to do cognitively new things. Reminiscent of the field theory that influenced both physics and psychology in the early twentieth century, Piaget thought that when accumulated knowledge reaches a critical mass a reorganization of knowledge becomes possible, resulting in the *emergence* (not an accidental use of this word) of some kinds of relational knowledge. Thus, from one developmental stage to the next one becomes not only more competent at practiced cognitive activity but actually capable of performing novel cognitive activity. Example: With practice at distinguishing the similarities and differences between or among objects and events, children get to be better and better at comparing. Once that accumulation of experience has reached a critical mass, and the need to do so is presented in a context that is important to the child, classification becomes possible as an emerging new cognitive ability. Successive experience then leads to class inclusion, with its concepts of superordinate and subordinate classes, and eventually to seriation (going inside a previously established class and arranging its items in serial order). This conception of understanding and novel activity emerging from a critical mass of experience is, in its broad outlines, suggestive of Rumbaugh's notion of emergents: innovative behavior based on the combination of prior experience and cognitive reorganization of knowledge elements.

So far so good, but one has to ask what energizes these cognitive developmental steps. In Piaget's case, it is disequilibrium. Discrepancy theories of motivation and perceptual cognition have been around for quite a long time: discrepancies between new information and stored information; discrepancies between arousing stimuli and one's optimal level of arousal. There is always in these conceptions an element of perceptual-cognitive incongruity, or a discrepancy between what is sensed and what is already known. Piaget proposed to resolve such discrepancies by the mechanism of equilibration, with its twin processes of assimilation and accommodation. If incoming information is familiar, it is assimilated and stored in preexisting cognitive compartments. If incoming information is so novel that it cannot be matched with or related to any already-stored elements, then either it must be reperceived (i.e., its meaning changed to be consonant with stored information) or the person's store of information must be changed. That is to say, one must change the meaning of the incoming information or change the person by revising the person's store of information. The energizing aspect of that notion derives, unfortunately, from the nineteenth century biology that sought steady states and quiescence: stop the disturbance of the system by reducing or eliminating the discrepancy.

A More Useful View of Reinforcement

To the extent that the role of reinforcement in behavior change (and adaptability) becomes conceptually strained, the concept needs to be defined differently. Rumbaugh et al. (2008) have defined reinforcement in a far less tautological way than have their behaviorist forebears. It is seen in terms of stimulus strength, salience, and the response-eliciting properties of adjacent (contiguous) stimuli. In its simplest form, salience refers to stimuli that have idiosyncratic value for the individual: what is important to me. The concept is reductionistic in the best sense, depending ultimately on brain structures and events. This also is a departure. Hebb (1949) complained that psychologists tend to view the brain as having all the finer structure of a bowl of porridge. This sad situation began to improve with the emergence of neuroscience, and inevitably cognitive neuroscience, in the 1970s and 1980s. What distinguishes rational behaviorism's concept of reinforcement is, first, that it is not intended to serve as a means of forming habits between stimuli and responses or to reduce need states or otherwise to increase the probability of behavior occurring in a given situation, but that reinforcers are to be understood in terms of what happens when two or more stimuli, sufficiently salient to draw attention, occur contiguously. When they do, they form amalgams in accordance with their relative strength, their defining attributes of salience, and their response-eliciting properties that reflect their natural and/or acquired characteristics. Thus, amalgams are produced by the species' neural systems and constitute units of experience that can then serve as a basis for knowledge, cognition, and intelligent adaptability. The concept of amalgams is reminiscent of Hebb's (1949) cell assemblies, neural organizations based on experience that upon reaching a critical accumulation may develop into phase sequences. Once strong and less-strong stimuli have become amalgamated, they come to share their response-eliciting properties in ways that are greatly influenced by the particular species' neural systems, which in turn determine the adaptability of the stimulus amalgams. Rumbaugh has suggested that amalgams can be likened to knowledge units, based on experience. They, in turn, are organized into templates via assimilation and accommodation (see discussion of Piaget) to the end of becoming organized for efficient use as needed by new situations (i.e., adaptable; perhaps "functionally autonomous" in the terms of Gordon Allport, 1937). Thus we have a salience theory substantially replacing more traditional concepts of reinforcement while helping to account for the accumulation of knowledge, amalgams representing aggregates of knowledge, behavioral (and cognitive) templates forming as amalgams are associated with each other in new ways (presumably as novel neural circuits are developed), and, from time to time, the emergence from those templates of novel, creative, innovative behavior (and cognition). One of the most salutary aspects of such a reconceptualization of reinforcement is the notion that disequilibrium, strong stimulation, activation, arousal, novel experience – all may have positive motivating effects rather than seeking reduction, and may lead to the continued seeking of new experience that may in turn provide the basis for the development and emergence of brand new abilities.

Rational Behaviorism and the Future

If it is true that

(a) the brain prefers to be active rather than to be a reactor only,
(b) adaptability is a more powerful evolutionary goal than is adaptation,
(c) organisms need to accumulate a knowledge/experience base before they can achieve more relational and abstract thinking,
(d) behavior, including cognitive activity, has to be energized,
(e) conditioning of reflexes or simple acts fails to explain innovative behavior,
(f) reinforcement as drive reduction leaves unexplained those aspects of behavior that are not associated with reduction of primary or secondary drives,
(g) it is difficult to locate the cultural survival/development value of reinforcement, and
(h) selection of goals and determination of the direction of behavior as well as its cessation are not adequately explained by S–R models,

then psychology is in great need of a conceptual structure than can take advantage of what is known on the basis of conditioning and learning theory and expand the knowledge base to encompass the emergence of innovative behavior – and cognition.

I focus here on only the two central concepts of rational behaviorism, *emergents* and *salience*. Of the two, the term emergents refers to the focal cognitive concept, whereas salience is the essential attentional, and ultimately motivational, concept. Salience is the property that determines what objects or events attract one's attention. The natural and/or acquired properties of two more co-occurring stimuli lead to the neural system's forming amalgams of them, consonant with the history of natural selection that shaped the constructive biases of its neural system's operations. The transactional nature of the two concepts leads us to a motivational theory of cognition (or a cognitive theory of motivation), in this case a substantial component of rational behaviorism (Rumbaugh, 2002; Rumbaugh et al., 2007; Rumbaugh & Washburn, 2003; Rumbaugh et al., 1996; Rumbaugh et al., 2008; see also Haywood, 2005, for a discussion of the transactional relations of intelligence, cognition, and motivation). It is the concept of emergents that helps one to bridge the gulf that otherwise separates conditioned responses from thinking and behavior.

Until the appearance of rational behaviorism, psychology had been best at explaining quantitative changes in behavior. Although many investigators have recognized the existence of creative, innovative behavior in human beings, few have been willing to acknowledge its existence in other animals, and even fewer have been able to explain how (and why) it occurs. Emergents constitute new, qualitatively different behavior that grows out of (a) cumulative experience with familiar situations, materials, and goals; (b) the amalgams (and subsequent templates) that grow up from the association of strong and less strong stimuli that have been experienced in the past in close association in space and/or time; (c) new demands for novel behavior; (d) the salience of the surrounding stimuli; and (e) the organism's evolutionary history and neural development.

What this set of concepts offers is a steady and logical progression, in a developmental sense, beginning with a reinterpretation of reinforcement, its reliance on the salience of stimuli, the contiguous relation of salient stimuli resulting in the brain's formation of amalgams. The brain then "sorts" amalgams according to need into neural templates that govern, lead to, make possible, and organize effective and novel behavior.

The brain's constant efforts to achieve optimal organization among the amalgams and templates leads to the occasional "nomination" of emergents and their more-or-less-successful application.

In the terms of this book, emergents may occur periodically in individuals, or in several members of a species, as isolated episodes that, demonstrating their value both for individuals and for groups (societies, cultures), get to be adopted and institutionalized as cultural phenomena. This progression offers a rich agenda for scholarly inquiry in the future. Motivational mechanisms deserve and require closer attention. The precise identification of emergents similarly needs research. We shall need exploration of the exact processes by which and over what periods of time individually experienced emergents pass into peer groups, clans, societies, and cultures. Similarly, the parameters of salience will require investigation: what stimulus conditions qualify as effective, how does salience shift within individuals over time and changing circumstances? The dependence of emergents on the salience of stimulus conditions is apparent, and of such importance that the two concepts are indeed symbiotic. The parameters of that relationship will have to be established. Similarly, we will need to know the precise conditions under which previously isolated knowledge structures form amalgams and amalgams form templates. These are enormous questions, but the guideposts are now present.

Acknowledgments

I owe Ed Wilson an enormous debt of gratitude for his generosity in sharing the recording of the Skinner/Wilson conversation with me years ago. The passage of time since it was recorded has only served to amplify its significance in the history of ideas, while dulling the sword of uninformed criticism leveled at both him and Skinner. I am also grateful for Ed's kindly challenge to produce this book and for his willingness to offer suggestions and editing along the way. Ed's contributions to science are profound and wide-ranging. Yet, they are outweighed by the warm dignity of his character and the intellectual encouragement he has offered so many. Many thanks are also due to Ed's longtime assistant, Kathleen Horton, who facilitated our correspondence and made certain I received appropriate attachments of Ed's relevant works-in-progress.

I am equally indebted to Julie (Skinner) Vargas and Ernest Vargas for their keen interest in this project. Both Julie and Ernest were extremely helpful in clarifying details, word choice, and concepts that, in their opinion, most appropriately represent the thinking of her father. The legacy of Skinner is assured with Julie's and Ernest's devoted work as key officers of and intellectual inspiration to the B.F. Skinner Foundation.

I offer my deepest appreciation to Duane Rumbaugh for his expert guidance. The legacy of Skinner, Wilson, and Rumbaugh in the history of twentieth century science will most certainly expand during this new century. Duane's professional relocation to the nearby Great Ape Trust of Iowa was truly a serendipitous and propitious event that provided me the unanticipated opportunity to learn from one of the world's most accomplished primatologists. His intellectual life has been one long and elegant argument supporting rational behaviorism and its essential concept of emergents. I consider myself genuinely privileged to count Duane as a colleague and close friend. His encouragement and support in completing this book made all the difference. Duane also introduced me to the important work of H. Carl Haywood of Vanderbilt University. It was quickly apparent that Carl would be the ideal scholar to provide an Epilogue for the book. I am grateful for his willingness to do so. His historical perspective and wisdom regarding indications for the future bring appropriate considerations to close the book.

Beyond the good fortune of acquiring a recording of the Skinner/Wilson conversation, and engaging Ed in a continuing conversation about his work, I have been the fortunate beneficiary of life circumstance that crossed paths with instrumental others. As a naïve and unknowing graduate student at The Ohio State University, I was witness to the rambling monologues from one of Harry Harlow's graduate students – Donald R. Meyer. Don's yearlong neuropsychology class and doctoral seminars were filled with frequent digressions, colorful anecdotes, elaborate blackboard drawings and empirical observations regarding the brain lesion studies he performed under Harlow's tutelage – lesions performed to determine the cortical location of learning sets. Those moments, combined with Don's "loosely compiled" notebooks, his willingness to share copies of his frequent handwritten letter correspondence with Donald Hebb – all fertilized by his generous willingness to simply talk – provided what I now understand to be a significant basis for my intellectual development in the behavioral sciences. Additional gratitude is due to others in that same Ohio State academic family – Delos D. Wickens and John O. Cooper, my behaviorist mentors, and Marlin Languis who always provided the right balance of challenge and support.

Special acknowledgement is due to my Central College faculty colleagues Mark Johnson and Keith Jones, and college trustee Rick Ryan for their interest in the project, their willingness to review working drafts and to offer suggestions for improvements. I also offer sincere thanks to the college president, David Roe, who offered me summertime flexibility to arrange the uninterrupted periods of time that were required to complete this project. Special mention and gratitude is due to Marilyn Vrban, my assistant, who keeps it all under control. A final thank you is owed to Janet Slobodien, life sciences editor with Springer. Her guidance and support along the way were very welcome.

Most of all, I am grateful to Ann – she understood. The project was completed because she endured many lost weekends and vacation days that allowed me the quality time necessary to focus. In addition, her comments were enormously helpful as I worked to achieve clarity and balance in the text. In the end, the project was completed because of her support and encouragement.

Glossary

Amalgams Duane Rumbaugh's term for a neuronal cell assembly that spontaneously forms out of a perceptual experience that is contiguous with environmental stimuli. The amalgam requires neither an unconditioned stimulus nor an operant reinforcer; it simply arises out of the contiguity of environmental event(s) and perceptual experience. Rumbaugh goes on to describe that the brain produces "emergents" as it necessarily works to resolve dissonance among amalgams (see *cell assembly*).

Behaviorism A concept directly related to John Watson's word choice in "Psychology as the Behaviorist Views It" (1913). American psychology began a significant departure from the considerations of covert mental processes as a result of Pavlov's work on classical conditioning in the late nineteenth century. "Mentalism" was increasingly rejected because instrumentation technology was not available that could provide for the empirical precision required to achieve validity and reliability. Behaviorists only accept data that are overt, easily observable, and quantifiable – behavior that is readily open to empirical manipulation and replication. This same behavioral approach is of primary importance to a biologist with serious interest in ethology and the evolution of social behavior – precise observation and careful quantification.

Canalization The developmental process that results in organisms' increasing predisposition to a within species phenotype (physical or behavioral) that is more narrowly represented by that which is most commonly observed as the within species norm. The term is derived from the fitting visual image the root word "canal" denotes – a narrow and focused passageway.

Cell assembly A term introduced by D.O. Hebb in 1949 (prior to sophisticated imaging technology) to provide a visual image of his proposed multineuron synaptic assembly that coded memories.

Classical conditioning The behavioral paradigm developed by Ivan Pavlov that describes the relationship among stimuli, their involuntarily caused responses, and previously neutral stimuli. When an unconditioned stimulus (UCS) is contiguously paired with a neutral stimulus (NS) on several occasions, the involuntary unconditioned response (UCR) will be elicited by the neutral stimulus in the absence of the unconditioned stimulus. Once this circumstance occurs, the neutral stimulus is then referred to as a conditioned stimulus and its result, a conditioned response (see *unconditioned stimulus, unconditioned response, neutral stimulus, conditioned stimulus, and conditioned response*).

Coevolution Represents an evolutionary change in a trait among individuals in one population that is responsive to a trait among individuals within another population that is then "followed by an evolutionary response by the second population to the change in the first" (Janzen, 1980). More specifically related to sociobiology, Wilson most frequently embeds the term within the phrase: gene-culture coevolution (see *gene-culture coevolution*). The term in that usage represents the mutually dependent reciprocal interaction of genes and culture that provides for the behavioral phenotype known as culture.

Conditioned response (CR) The result of contiguously pairing a neutral stimulus (NS) with an unconditioned stimulus (UCS), initially resulting in an involuntary unconditioned response (UCR). When the neutral stimulus is paired on enough occasions with the unconditioned stimulus, the neutral stimulus may be presented in isolation and still elicit the previously unconditioned response. The point at which this "learned" behavior is observed, the formerly neutral stimulus is referred to as a conditioned response (CR) (see *classical conditioning*).

Conditioned stimulus (CS) Once a neutral stimulus (NS) is capable of eliciting the unconditioned response (UCR) in the absence of the unconditioned stimulus (UCS), the neutral stimulus is referred to as a conditioned stimulus (see *classical conditioning*).

Continuous schedule of reinforcement Skinner's operant conditioning paradigm is defined by the naturally occurring or artificially structured consequences to the emitted free operants (behavior) of organisms. Continuous reinforcement rarely happens in the natural environment, yet is commonly the means by which operant behavior is initially conditioned in a laboratory or clinical setting in order to assure behavioral acquisition. In the example, where an experimenter is conditioning a laboratory rat to press a bar for food, continuous reinforcement means a food pellet is released each time the rat presses the bar.

Cue/discriminative stimulus (S^D or S^Δ) Free operants are never "caused" in the truest sense of the word's meaning; instead, they are emitted by the organism. However, there exists a category of stimuli that make it more (S^D) or less (S^Δ) likely that the organism will emit the behavior because they signal the prospect that reinforcement is at hand (S^D) if the behavior is produced or punishment (S^Δ) will result if the behavior is emitted. This class of stimuli is defined as discriminative stimuli or cues in the operant conditioning paradigm (see *modeling* and *priming*).

Culture "Information capable of affecting individual behavior acquired from other members of the species by teaching, imitation, and other forms of social transmission" (Boyd and Richerson, 2005). Human culture is characterized by traits common to all members of the species (culturgens) and by idiosyncratic traits shared within particular social groups (Lumsden and Wilson, 1981).

Culturgen The term coined by E.O. Wilson to embody the array of transmissible behavior, mental constructs, and artifacts that represent the basic units of culture. The term is derived from *cultura* – Latin for culture and *gen* – Latin for produce. Terms coined by others to denote similar concepts include: mnemotype, idea, idene, meme, and sociogene. The most precise use of the term is not wholly dissimilar to Dawkins' *meme*, and is used to define the generators of culture. Years later Wilson abdicated the term *culturgen* in favor of Dawkins' *meme*, which had achieved far more regular usage in the academic literature.

Differential reinforcement The procedure used to operantly condition a sophisticated behavior not within an organism's current repertoire of behavior, yet one which the organism is physically capable of producing. The process is initiated by carefully identifying the essential components of the desired terminal behavior and selectively (differentially) reinforcing the most basic element of that behavior emitted by the organism. As that basic element of the terminal behavior becomes conditioned, the behavioral contingency for reinforcement will be increased differentially to successively approximate (see *successive approximation*) the terminal behavior. Behavioral training of animals is accomplished by this procedure. Skinner enjoyed demonstrating how he could operantly condition a parrot to play ping-pong or basketball by utilizing differential reinforcement of successive approximations of the terminal behavior.

Echo neuron A category of mirror neurons (see *mirror neuron*) that function within the auditory system and are considered to have a significant impact on the imitative functions related to language acquisition.

Emergents A new category of behavior pattern that adds to the existing respondent and operant behavior. Emergents arise spontaneously as a result of the synergistic relationship among cell assemblies and are not due to an explicit stimulus or perception. They are noted for being synergistic, integrative, and clever. They manifest themselves as new capabilities that can't be explained as responses or behavior altered by basic conditioning procedures.

Epigenesis The process of interaction between genes and environment that leads to phenotypic expression in organisms. Wilson focuses his usage of the term on the process of gene-culture coevolution that results in the phenotypic expression of behaviors that are typically described as social and cultural behaviors.

Epigenetic rule Rules describing the regularities that emerge during the interaction of genes with the environment (epigenesis) that "channel the development of an anatomical, physiological, cognitive, or behavioral trait in a particular direction. Epigenetic rules are ultimately genetic in basis, in the sense that their particular nature depends on the DNA developmental blueprint. Some epigenetic rules are inflexible, with the final phenotype being buffered from all but the most drastic environmental changes. Others permit a flexible response to the environment; yet even these may be invariant, in the sense that each possible response in the array is matched to one environmental cue or set of cues through the operation of special control mechanisms. In cognitive development, the epigenetic rules are expressed in any one of the many processes of perception and cognition to influence the form of learning and the transmission of culturgens" (Lumsden and Wilson, 1981). Increasingly, evolutionary psychology is suggesting (as an example) the prospect that humans' predisposition to religious behavior is responsive to epigenetic rules that guide us to the culturgens of specific religions available in our immediate environment.

Ethology The biological study of animal behavioral patterns in natural settings. Most typically, a specific behavior is studied across species in order to analyze adaptation and patterns of evolution.

Euculture Lumsden and Wilson (1981) use the term to embody the most advanced form of culture. They suggest that this level of culture is uniquely occupied by humankind because of our capacity not only to teach and learn (shared with other species), but also to conceptualize the content of teaching and learning into "concrete entities that can be more readily labeled by symbols and handled by language."

Evolutionary psychology The approach in psychology that is based on the assumption that all psychological functions are biological adaptations based in Darwinian selection theory. Thus, evolutionary principles are applied in describing near-universal behavioral predispositions and a wide-range behavioral repertoires. Although mostly limited to humans, evolutionary psychologists also base much of their work on other organisms.

Evo devo Represents the shorthand version of the term: *evolutionary developmental biology*. The essential basis for the field is the investigation of genetic and developmental mechanisms that function during embryonic development to guide how organismal morphology (form) can be modified by evolutionary alterations in the developmental process. Current research is focused on how even the most subtle changes in that developmental process produced by so-called "tool-kit genes" and "genetic switches" can alter the developmental process enough to initiate speciation (see *tool kit genes*).

Free operant behavior (see *operant behavior*).

Gene-culture coevolution The codependent evolutionary process of genes and culture. "More precisely, any change in gene frequencies that alters culturgen frequencies in such a way that the culturgen changes alter the gene frequencies as well" (Lumsden and Wilson, 1981). (see *culturgen*).

Generalized (secondary) reinforcer (see *positive reinforcement, primary reinforcement,* and *secondary reinforcement*) A positive reinforcer that achieves its value by early pairing with a primary reinforcer. The generalized nature of the reinforcer means that it can be exchanged for a variety of desired consequences, including primary reinforcers. The most typical example is money. The positive attributes of money must initially be learned before having any reinforcing value; thus, making it a secondary reinforcer. It can be used in exchange for all manner of things by choice of the individual receiving it as a consequence to producing a desirable behavior.

Genotype The genetic composition of an individual organism (see *phenotype*).

Group selection The natural selection process that favors characteristics enhancing the prospects for survival of one group relative to another group. If group selection acts effectively, the group may evolve into an adaptive unit and be open to study in the same way individuals can be studied (see *multilevel selection*).

Intermittent schedule of reinforcement (see *continuous reinforcement*). Reinforcement that, by design, is not made available every time an organism produces a desired behavior. A strategic and gradual increase in the contingency to achieve reinforcement will typically produce behavioral longevity in the organism emitting the behavior without producing an "expectation" for frequent reinforcement (see *resistance to extinction*).

Kin selection "The change in gene frequencies due to one or more individuals favoring or disfavoring the survival and reproduction of relatives (other than offspring) who possess the same genes by common descent." (Lumsden and Wilson, 1981). (see *multilevel selection*).

Lamarck/Lamarckian The theory of evolution developed by Jean Baptiste de Lamarck (1744–1829). The theory identifies that individual characteristics acquired by the activity of an organism are passed directly to offspring in the next generation.

Law of effect One of two primary laws of behavior developed by Edward Thorndike. The law identifies that organisms will tend to continue producing behaviors that result in a favorable consequence (effect) and will stop producing behaviors that result in unpleasant consequences.

Law of multiple responses Thorndike's second primary law that suggests organisms will engage in a variety of behaviors until one achieves a pleasant consequence – a kind of behavioral trial and error of sorts.

Learning set A concept identified by Harry Harlow (Harlow, 1949) to describe his observations of animal behavior following cumulative experience with a significant number of problems comprised of a single class. Harlow suggested that a more Gestalt learning model prevailed and subsequent novel problems within that same class would be solved insightfully and "supplanted by an orientation toward task accomplishment via creative actions, afforded by emergent processes that generate new options of choice both behaviorally and intellectually."

Meme Coined by Richard Dawkins, the term first appeared in *The Selfish Gene* (1976). Dawkins (1982) identifies a meme as "a unit of cultural inheritance naturally selected by virtue of its 'phenotypic' consequences on its own survival and replication." He adds that it is also a "unit of information residing in the brain." Memes have the capacity to advance themselves in the meme pool by their capacity to leap from brain to brain in a process that is best characterized by imitation. Since its initial introduction, it has achieved widespread acceptance as the term of choice to represent units of culture.

Mirror neuron A group of neurons that actively fire when an organism is producing behavior or when the organism observes the behavior being produced by another organism. Mirror neurons provide a significant neural substrate for imitation and have obvious implications to research efforts on cued motor behavior and language acquisition (see *echo neuron*) in humans.

Modeling An intentional demonstration of a behavior to an observer/learner that typically includes an ambition that the observer/learner will attempt to emit the behavior on their own. In a designed operant environment, the modeler/teacher will reinforce the modeled behavior, aspiring to increase the likelihood the behavior will recur (see *cue* and *priming*).

Multilevel selection Natural selection, as the driving force of adaptation, can occur within multiple levels of the biological hierarchy. Thus, natural selection can operate "among individuals within a single population and then frame-shift upward to selection among groups in a metapopulation" (Wilson, 1997). Indeed, the individual organism itself is on this multilevel continuum, in that it is "a higher level unit – a group of genes." The concept of group selection resides within the larger framework of multilevel selection (see *group selection*).

Negative punishment (S^P–) The removal of a stimulus an individual would like to retain as a consequence to their emitting an undesired behavior. A fine imposed following a speeding violation and imprisonment following a more serious crime are both examples of stimuli we all wish to maintain – loss of money and loss of freedom.

Negative reinforcement (S^R–) Any consequence to a behavior that is likely to increase its probability of recurrence is characterized as a reinforcer. The increase in behavior is brought about by the prospect a negative stimulus will be presented if the behavior is not emitted. Most behaviorists suggest that negative reinforcement is the most common basis for the increased probability of appropriate behavior. The majority of humans behave (R) in socially appropriate ways in order to avoid the prospect of being rejected by peers or punished by eternal damnation. The punishing consequent is not played out – it is the prospect for the punishing consequent. Children will pick up their clothes and make their bed in order to avoid the prospect that TV time will be reduced by a parent.

Neoteny The tendency among primates and humans to developmentally retain juvenile features in physical attributes for extended periods of time – oftentimes, well into adulthood. Developmental biologists view this attribute as a significant mechanism in achieving neural plasticity for decades among humans, allowing for the gene-culture interaction so critical to the emergence of human behavioral patterns.

Neutral stimulus The stimulus within the classical conditioning paradigm that has no eliciting relationship to the caused unconditioned response (UCR) prior to its contiguous pairing with the causal unconditioned stimulus (UCS). Once the neutral stimulus (NS) is consecutively paired with the UCS in presentation several times, the NS may be presented in the absence of the UCS and still elicit the initial response. Once this conditioning is demonstrated, the NS is referred to as a conditioned stimulus (CS) that elicits a conditioned response (CR).

Ontogeny The physical development of an organism from embryo through adulthood.

Operant behavior Coined by B.F. Skinner to represent all uncaused behavior emitted by organisms. The environmental consequences to these "free" operants will increase or decrease the likelihood the organism will emit the operant behavior again.

Operant conditioning The behavioral conditioning paradigm developed by B.F. Skinner that scientifically manipulates the consequences to free operant behavior intended to increase or decrease the likelihood the behavior will recur. Reinforcers increase the prospects for a behavior to occur again, while punishers decrease the probability the behavior will be emitted again (see *operant, positive reinforcement, negative reinforcement, positive punishment, and negative punishment*).

Organism The conventional (although incorrect) use of this term is to represent any living creature capable of emitting free operant behavior and, therefore, susceptible to the imperatives of operant conditioning. Its correct usage is to represent anything that is living, including plants or other organisms that are incapable of emitting free operant behavior.

Phenotype The physical and behavioral characteristics of an individual that result from the interaction between genotype and environmental factors (see *genotype and epigenesis*).

Phylogeny The evolutionary history of a species' lineage as it emerges over time. Phylogeny attempts to reconstruct the genotypic and phenotypic sequence of an organism's evolutionary development.

Plasticity A commonly used term in neuroscience to indicate the phenotypic malleability of an organism as it develops over time. Organisms having more significant periods during which they maintain neotenous (see *neoteny*) attributes sustain a greater measure of developmental plasticity. Plasticity in neurological development enables an organism to be more behaviorally adaptive to environmental pressures without the burden of maladaptive stereotypic responses "caused" by hardwired brain organization. The magnitude of neurological plasticity is positively correlated with the ratio of cortex to body mass.

Positive punishment (S^P+) An unpleasant stimulus presented as a consequence to an undesirable behavior designed to stop or diminish the behavior. The common childhood spanking is the best example.

Positive reinforcement (S^R+) A pleasant stimulus presented as a consequence to a desirable behavior that increases the likelihood the behavior will be emitted again. Skinner's work unequivocally demonstrated that reinforcement was most effective when made available contiguous with emitting the behavior. Verbal praise, a hug, a token for exchange, or TV time are common childhood examples. The most typical positive reinforcement for adults is the paycheck.

Primary reinforcer Positive reinforcement can be classified as primary or secondary (see *secondary reinforcer*). A primary reinforcer achieves its reinforcing value because it meets a basic survival need – it achieves its relative effectiveness by creating a state of deprivation. Food, water, air, and sleep are examples. Skinner was adamantly opposed on ethical grounds to utilize primary reinforcers as a means to operantly condition human behavior.

Priming An operant term that is often used in the context of discriminative stimuli or cues that is meant to suggest a conscious arrangement of stimuli in an environment to establish inevitability that the organism will emit the desired behavior (see *cue*, *modeling,* and *discriminative stimuli*).

Proximate cause The immediate functional elements of the environment that drive behavior. "Proximate causes govern the responses of the individual (and his organs) to immediate factors of the environment" (Mayr, 1961) (see *ultimate cause*).

Rational behaviorism A concept advanced by Duane Rumbaugh and his colleagues that provides for a "trichotomous classification of behavior that recognizes and adds to Skinner's (1938) distinction between respondent and operant conditioning" (Rumbaugh et al., 1996). Rumbaugh and his colleagues (1996) suggested that prevailing twentieth century behavioral concepts of respondents and operants didn't explain the *thought* that was apparent in Harlow's "learning set" work, which led them to propose an entirely new class of behavior – *emergents*. This third class of behavior serves as the basis for rational behaviorism, which "allows for the emergence of new behavior generated by cognitive operations of the primate brain . . . reflect(ing) the natural operations of the brain as being comprised of keen pattern-detection and synthesizing systems" (Rumbaugh, 2002) (see *emergents*).

Reductionism The attempt to explain all biological and behavioral processes by utilizing the mechanistic principles of physical laws. A reductionist approach attempts to break down complex systems into component parts and then describe the entire system on the basis of the physical properties of each part. The term is oftentimes used as a means to criticize the work of the scientific community as oversimplifying complex systems that are greater than the sum of their parts (i.e., neuroscientists who suggest that human consciousness can be explained as via mechanistic brain processes are accused of being reductionistic).

Reify/Reification Humankind's unique mental activity that fully separates us from other advanced species and makes us the only known eucultural species. Mostly based in structures of language and symbolic visual representations, reification enables humans to produce concepts and classification schemes that make meaning of the world we experience. Reified concepts and symbols take on a more "concrete" form as a thing that can be transmitted from one person to another via the mechanisms of culture. The word is based on the Latin root – *res*, meaning thing.

Reinforcer (reinforcing stimuli) Any stimulus following a behavior that increases the likelihood the behavior will be emitted again (see *positive reinforcement*).

Respondent behavior This phrase is most often shortened to "respondent" in much the same fashion as operant behavior is shortened to "operant." A respondent (or respondent behavior) is the behavior elicited by a specific stimulus presented to an organism (see *respondent conditioning*).

Respondent conditioning A term sometimes used in place of *classical conditioning*. The term is most frequently encountered in texts more focused on Skinner's operant conditioning paradigm, in that Skinner referred to classical conditioning as respondent conditioning. He did so because an organism is "responding" to an environmental antecedent that causes the response (see Classical Conditioning).

Sociobiology "The systematic study of the biological basis of all forms of social behavior, including sexual behavior and parent-offspring interaction, in all kinds of organisms" (Lumsden and Wilson, 1981). The term was originally coined and used by E.O. Wilson in *The Insect Societies* (1971).

Secondary reinforcer Positive reinforcement can be classified as primary or secondary (see *primary reinforcer*). A secondary reinforcer achieves its reinforcing value only after the organism "learns" its positive attributes by early pairing with a primary reinforcer. That is, the verbal praise that is commonly paired with the contact comfort (primary reinforcer) with a young child can quickly become a secondary reinforcer and will function as a reinforcer in the absence of a primary reinforcer (see *positive reinforcement, primary reinforcer, and generalized secondary reinforcer*).

Shaping (operant shaping) An operant conditioning technique that conditions complex behavior not within an organism's normal repertoire of behavior. Operant shaping toward a desired target behavior (sometimes referred to as *terminal behavior*) results from strategically increasing the behavioral expectation for reinforcement to successively approximate the target behavior (see *successive approximation*). Ultimately, the final behavior can be achieved with patience and vigilance to the technology of operant shaping. That is, Lassie must be first conditioned to attend to her trainer before she is reinforced for approaching a light switch on verbal command. There is considerable time and many intervening approximations that must be shaped before Lassie stands against the wall and flips on the light switch with her extended paw

Successive approximation The strategic and gradual increase of behavioral expectation toward an identified target behavior required of an organism to achieve reinforcement.

Tool kit genes The concept identified to describe the near universal building-block genes that provide for the physical construction of an organism. Tool-kit genes are responsible for the timing and expression of genotype into phenotype. Various animal architectures (i.e., the eye) are produced by applying the same genetic tools in different ways (see *evo devo*).

Ultimate cause Evolutionary mechanisms that drive behavior. "Ultimate causes are responsible for the evolution of the particular DNA code of information with which every individual of every species is endowed" (Mayr, 1961) (see *proximate cause*).

Unconditioned response (UCR) The involuntary response elicited by an unconditioned stimulus. That is, light (UCS) will always cause the pupil to constrict (UCR).

Unconditioned stimulus (UCS) A stimulus that elicits an involuntary/caused response (see *unconditioned response*). The stimuli act on a fairly small class of behavior most often limited to those related to smooth muscle and glandular mechanisms. That is, a strong puff of air to the eye (UCS) will always cause a blink of the eyelid (UCR) (see *classical conditioning*).

Vicarious reinforcement The phenomenon whereby an individual observing another individual being reinforced or punished for a behavior can lead to increasing or decreasing the likelihood that the observer will emit the same behavior. That is, observing someone earn praise for losing weight can lead to the observer initiating an exercise program or a diet.

Bibliography

Allport, G. (1937). *Personality: A Psychological Interpretation.* New York: Holt, Rinehart, & Winston.

Alcock, J. (1975). *Animal Behavior: An Evolutionary Approach.* Sunderland, MA: Sinauer Associates.

Amundson, R. (2005). *The Changing Role of the Embryo in Evolutionary Thought: Structure and Synthesis.* New York: Cambridge University Press.

Aunger, R. (ed.). (2000). *Darwinizing Culture: The Status of Memetics as a Science.* New York: Oxford University Press.

Barash, D. (2005). "B.F. Skinner, Revisited." *The Chronicle of Higher Education*, 51, 30, B-10.

Barkow, J.H. Cosmides, L., and Tooby, J. (1992). *The Adapted Mind: Evolutionary Psychology and the Generation of Culture.* New York: Oxford University Press.

Blackmore, S. (1999). *The Meme Machine.* New York: Oxford University Press.

Blackmore, S. (2005). "Evidence for Memetic Drive?" S. Hurley and N. Chater, (eds.), *Perspectives on Imitation: From Neuroscience to Social Science. v. 1.* Cambridge, MA: MIT Press.

Boring, E.G. (1950). *A History of Experimental Psychology.* 2nd Ed. New York: Appleton-Century-Crofts.

Boyd, R. and Richerson, P.J. *The Origin and Evolution of Cultures.* (2005). New York: Oxford University Press.

Brown, J. (1975). *The Evolution of Behavior.* New York: W.W. Norton.

Bjorklund, D.F. and Pellegrini, A.D. (2000). "Child Development and Evolutionary Psychology." *Child Development.* 71, 1687–1708.

Bjorklund, D.F. and Pellegrini, A.D. (2002). *The Origins of Human Nature: Evolutionary Developmental Psychology.* Washington, DC: American Psychological Association.

Buller, D.J. (2005). *Adapting Minds: Evolutionary Psychology and the Persistent Quest for Human Nature.* Cambridge, MA: Bradford Books/MIT Press.

Carroll, S.B. (2005). *Endless Forms Most Beautiful: The New Science of Evo Devo.* New York: W.W. Norton & Co.

Darwin, C. (1859). *On the Origin of Species by Means of Natural Selection, or the Preservation of Favoured Races in the Struggle for Life.* London: John Murray.

Darwin, C. (1859). *The Origin of Species.* New York: Modern Library (1993 edition).

Darwin, C. (1871). *The Descent of Man.* New York: Penguin Classics (2004 edition).

Darwin, C. (1872). *The Expression of the Emotions in Man and Animals.* New York: Oxford University Press (2002 edition).

Darwin, C. (1929). *The Autobiography of Charles Darwin.* Cambridge, UK: Icon Books Ltd. (2003 edition).

Dawkins, R. (1976). *The Selfish Gene.* New York: Oxford University Press.

Dawkins, R. (1982). *The Extended Phenotype: The Gene as Unit of Selection.* San Francisco: W.H. Freeman.

Dawkins, R. (1987). *The Blind Watchmaker.* New York: W.W. Norton & Co.

Dobzhansky, T. (1937). *Genetics and the Origin of Species.* New York: Columbia University Press.

Dobzhansky, T. (1973). "Nothing in Biology Makes Sense, Except in the Light of Evolution." *American Biology Teacher.* 35, 125–29.

Donahoe, J.W. (1984). "Skinner – The Darwin of Ontogeny?" *The Behavioral and Brain Sciences*, 7(4), 487–88.

Durant, J. (1980). "How Evolution Became a Scientific Myth." *New Scientist*, September 11, 765.

Eibl-Eibesfeldt, I. (1975). *Ethology: The Biology of Behavior.* New York: Holt, Rinehart and Winston.

Ferster, C.B. and Skinner, B.F. (1957). *Schedules of Reinforcement.* New York: Appleton Century Crofts.

Geary, D.C. (2006). "Evolutionary Developmental Psychology: Current Status and Future Directions." *Developmental Review*, 26, 113–119.

Grafen, A. and Ridley, M. (eds.). (2006). *Richard Dawkins: How a Scientist Changed the Way We Think*. New York: Oxford University Press.

Gould, S.J. (2002). *The Structure of Evolutionary Theory.* Cambridge, MA: The Belknap Press of Harvard University Press.

Granit, R. (1947). *Sensory Mechanisms of the Retina*. London: Oxford University Press; New York: Hafner (1963).

Granit, R. (1977). *The Purposive Brain*. Cambridge, MA: MIT Press.

Guthrie, E.R. (1952). *The Psychology of Learning: Revised Edition*. Boston: Harper Brothers.

Haeckel, E. (1894). *Die Systematische Phylogenie*. Berlin: Georg Reimer.

Haidt, J. (2007). "The New Synthesis in Moral Psychology." *Science.* 316, 998–1002.

Hamer, D. (2004). *The GOD Gene.* New York: Doubleday.

Hamilton, W. (1964). "The Genetical Evolution of Social Behavior, I and II." *Journal of Theoretical Biology.* 7, 1–52.

Harlow, H.F., Harlow, M.K, and Meyers, D.R. (1950). "Learning Motivated by a Manipulation Drive." *Journal of Experimental Psychology*. 40(2), 228–34.

Harlow, H.F. (1949). "The Formation of Learning Sets." *Psychological Review*, 56, 51–65.

Haywood, H.C. (2005). "A Transactional Perspective on Mental Retardation." In H.N. Switzky (Ed.), *Mental Retardation, Personality, and Motivational Systems: International Review of Research in Mental Retardation*, 31, 289–314. New York and Amsterdam: Elsevier/Academic Press.

Hebb, D.O. (1949). *The Organization of Behavior: A Neuropsychological Theory*. New York: John Wiley and Sons.

Henriques, G. (2008). "The Problem of Psychology and the Integration of Human Knowledge: Contrasting Wilson's Consilience with the Tree of Knowledge System." Theory and Psychology, 18(6), 731–755.

Hurley, S. and Chater, N. (eds.). *Perspectives on Imitation*. v. 1 & 2. Cambridge, MA: MIT Press.

Huxley, J. (1942). *Evolution: The Modern Synthesis.* London: Allen and Unwin.

Karpov, Y.V. (2005). *The neoVygotskian Perspective on Child Development*. New York: Cambridge University Press.

Karpov, Y.V. & Haywood, H.C. (1998). Two Ways to Elaborate Vygotsky's Concept of Mediation: Implications for Education. *American Psychologist*, 33(1), 27–36.

Koffka, K. (1935). *Principles of Gestalt Psychology*. New York: Harcourt, Brace.

Kuhn, T. (1970). *The Structure of Scientific Revolutions.* Chicago: The University of Chicago Press.

Lehrer, J. (2007). *Proust Was a Neuroscientist.* Boston: Houghton Mifflin Company.

Lickliter, R. (1996). "Structured Organisms and Structured Environments: Development Systems and the Construction of Learning Capacities. J. Valsinerand and H. Voss (eds.), *The Structure of Learning Processes*, 86–107. Norwood, NJ: Ablex.

Loeb, J. (1900). *Comparative Physiology of the Brain and Comparative Psychology.* New York: Putnam.

Lucretius, (50 BCE) *On the Nature of Things.* Translation by Martin Ferguson Smith, 2001, Indianapolis: Hackett Publishing Company.

Lumsden, C. and Wilson, E.O. (1981). *Genes, Mind and Culture*. Cambridge, MA: Harvard University Press.

Lumsden, C. and Wilson, E.O. (1983). *Promethean Fire*. Cambridge, MA: Harvard University

Press.
Mach, E. (1915). *The Science of Mechanics: A Critical and Historical Account of its Development.* Chicago: Open Court Press.
MacLean, P.D. (1949). "Psychosomatic Disease and the 'Visceral Brain': Recent Developments Bearing on the Papez Theory of Emotion." *Psychosomatic Medicine*, 11, 338–353.
MacLean, P.D. (1952). "Some Psychiatric Implications of Physiological Studies on Frontotemporal Portion of Limbic System." *Electroencephalography and Clinical Neurophysiology*, 4, 407–418.
MacLean, P.D. (1989). *The Triune Brain in Evolution.* New York: Plenum Press.
Maestripieri, D. and Roney, J.R. (2006) "Evolutionary Developmental Psychology: Contributions from Comparative Research with Nonhuman Primates." *Developmental Review*, 26, 120–137.
Mayr, E. (1942). *Systematics and the Origin of Species from the Viewpoint of a Zoologist.* Cambridge: Harvard University Press.
Mayr, E. (1961). "Cause and Effect in Biology." *Science*, 134, 1501–1506.
Mayr, E. (2001). *What Evolution Is.* New York: Basic Books.
Minelli, A. (2006). "The Roots of Evo-Devo." *Heredity*, 96, 419–420.
Müller, J. (1837–1840). *Handbuch der Physiologie des Menschen für Vorlesungen.* 2 Vols. Coblenz (Germany): Verlag von J. Hölscher.
Pavlov, I.P. (1927). *Conditioned Reflexes: An Investigation of the Physiological Activity of the Cerebral Cortex.* London: Oxford University Press.
Pavlov, I.P. (1928). *Lectures on Conditioned Reflexes.* New York: International.
Peterson, N. (1960). "Control of Behavior by Presentation of an Imprinted Stimulus." *Science*, 132(3437), 1395.
Piaget, J. (1950). *Introduction à l'Épistémologie Génétique.* Paris: Presses Universitaires de France.
Piaget, J. (1952). *The Origins of Intelligence in Children.* New York and Paris: International Universities Press.
Pinker, S. (2002). *The Blank Slate: The Modern Denial of Human Nature.* New York: Viking.
Plotkin, H. (2000). "Culture and Psychological Mechanisms." R. Aunger (ed.), *Darwinizing Culture: The Status of Memetics as a Science.* New York: Oxford University Press.
Rand, K.L. and Ilardi, S.S. (2005). "Toward a Consilient Science of Psychology." *Journal of Clinical Psychology*, 61(1), 7–20.
Rentfrow, P.J., Gosling, S.D., & Potter, J. (2008). A Theory of the Emergence, Persistence, and Expression of Geographic Variation in Psychological Characteristics. *Perspectives on Psychological Science*, 3(5), 339–369.
Richerson, P.J. and Boyd, R. (2005). *Not by Genes Alone: How Culture Transformed Human Evolution.* Chicago: University of Chicago Press.
Rizzolatti, G. and Craighero, L. (2004). "The Mirror-Neuron System." *Annual Review of Neuroscience*, 27, 169–192.
Rizzolatti, G. (2005). "The Mirror Neuron System and Imitation." S. Hurley and N. Chater, (eds.), *Perspectives on Imitation: From Neuroscience to Social Science. v 1.* Cambridge, MA: MIT Press.
Roberts, J.S. (2005). "Rooting for Evo-devo: a Review of *The Changing Role of the Embryo in Evolutionary Thought: Roots of Evo-Devo.*" *Evolution and Development*, 7, 647–648.
Rogers, C.R. and Skinner, B.F. (1956). "Some Issues Concerning the Control of Human Behavior: A Symposium." *Science*, 124, 1057–66.
Rumbaugh, D.M. (1984). "Perspectives by Consequences: A Commentary on Skinner's Selection by Consequences." *The Behavioral and Brain Sciences*, 7(4), 496–97.
Rumbaugh, D.M. (1995). "Emergence of Relations and the Essence of Learning: A Review of Sidman's Equivalence Relations and Behavior: A Research Story." *The Behavior Analyst*, 18, 367–375.
Rumbaugh, D.M. (1997). "The Psychology of Harry Harlow: A Bridge from Radical to Rational Behaviorism." *Philosophical Psychology.* 10(2), 197–210.
Rumbaugh, D.M. (2002). "Emergents and Rational Behaviorism." *Eye on Psi Chi.* 6, 8–14.

Rumbaugh, D.M. (2008a). Personal Communication: 5/28/08.

Rumbaugh, D.M. (2008b). Personal Communication: 6/2/08.

Rumbaugh, D.M., King, J.E., Beran, M.J., Washburn, D.A. and Gould, K.L. (2007). "A Salience Theory of Learning and Behavior: with Perspectives on Neurobiology and Cognition." *International Journal of Primatology*. 28(5), 973–96.

Rumbaugh, D.M. and Washburn, D.A. (2003). *Intelligence of Apes and Other Rational Beings.* New Haven: Yale University Press.

Rumbaugh, D.M., Washburn, D.A. and Hillix, W.A. (1996). "Respondents, Operants and *Emergents:* Toward an Integrated Perspective on Behavior." K. Pribram and J. King (eds.), *Learning as a Self-Organizing Process.* (pp. 57–73). Hillsdale: Lawrence Erlbaum Associates.

Rumbaugh, D.M., Washburn, D.A., King, J.E., Beran, M.J., Gould, K., & Savage- Rumbaugh, S. (2008). "Why Some Apes Imitate and/or Emulate Observed Behavior and Others Do Not: Fact, Theory, and Implications for Our Kind." *Journal of Cognitive Education and Psychology* (online), 7, 101–110. www.iacep.coged.org/journal

Segerstråle, U. (2000). *Defenders of the Truth: The Battle for Science in the Sociobiology Debate and Beyond.* New York: Oxford University Press.

Skinner, B.F. (1935). "On the Generic Nature of the Concepts of Stimulus and Response." *The Journal of General Psychology*. 12, 40–65.

Skinner, B.F. (1938). *The Behavior of Organisms: An Experimental Analysis.* New York: Appleton-Century.

Skinner, B.F. (1948). *Walden Two*. New York: Macmillan.

Skinner, B.F. (1953). *Science and Human Behavior*. New York: The Free Press.

Skinner, B.F. (1957). *Verbal Behavior*. New York: Appleton.

Skinner, B.F. (1961). "Evolution and Man's Progress." *Daedalus*, 90, 534–46.

Skinner, B.F. (1963). "Behaviorism at Fifty." *Science*, May 31(140), 951–958.

Skinner, B.F. (1966). "The Phylogeny and Ontogeny of Behavior." *Science*, 153(3741), 1205.

Skinner, B.F. (1971) *Beyond Freedom and Dignity*. New York: Alfred A. Knopf.

Skinner, B.F. (1981). "Selection by Consequences." *Science*, 213, 501–504.

Skinner, B.F. (1984). "Selection by Consequences." *Behavioral and Brain Sciences*, 7(4), 477–510. (Target article and commentary)

Skinner, B.F. (1989). *Recent Issues in the Analysis of Behavior.* Columbus: Merrill Publishing Company.

Skinner, B.F. (1990). "Can Psychology Be a Science of Mind?" *American Psychologist*, 45, 1206–1210.

Skinner, B.F. (1999). *Cumulative Record: Definitive Edition.* Acton, Massachusetts: Copley Publishing Group.

Sober, E. and Wilson, D.S. (1998). *Unto Others: The Evolution and Psychology of Unselfish Behavior.* Cambridge, MA: Harvard University Press.

Spinoza, B. (1677). *Ethics – The Collected Works of Spinoza* (E.M. Curley, ed.), 1985 edition. Princeton, NJ: Princeton University Press.

Thompson, N.S. (2000). "Shifting the Natural Selection Metaphor to the Group Level." *Behavior and Philosophy*, 28, 83–101.

Thorndike, E.L. (1898). "Animal Intelligence: an Experimental Study of Associative Processes in Animals." *Psychological Review, Monograph Supplement*, 2(8, 1), 16.

Thorndike, E.L. (1913). *Educational Psychology: The Psychology of Learning.* New York: Teachers College Press.

Thorndike, E.L. (1931). *Human Learning.* New York: Century.

Tolman, E.C. (1932). *Purposive Behavior in Animals and Men*. New York: Century.

Tolman, E.C. (1958). *Behavior and Psychological Man: Essays in Motivation and Learning.* Berkeley, CA: University of California Press.

Tooby, J. and Cosmides. L. (1992). "The Psychological Foundations of Culture.", L. Cosmides and J. Tooby (eds.), *The Adapted Mind: Evolutionary Psychology and the Generation of Culture* 19–136. New York: Oxford University Press.

Trivers, R.L. (1971). "The Evolution of Reciprocal Altruism." *The Quarterly Review of Biology.* 46, 35–57.
Vargus, E. (2008). Personal Communication: 9/12/08.
Vygotsky, L.S. (1978). *Mind in Society: The Development of Higher Psychological Processes* (Michael Cole, ed.). Cambridge, MA: Harvard University Press.
Vygotsky, L.S. (1986). *Thought and Language* (Trans. and Ed. by Alex Kozulin). Cambridge, MA: MIT Press.
Washburn, D.A. (2006). "The Perception of Emergents." D. Washburn (ed.), *Primate Perspectives on Behavior and Cognition,* 109–23. Washington: American Psychological Association Press.
Washburn, D.A. (ed.). (2007). *Primate Perspectives on Behavior and Cognition.* Washington: American Psychological Association Press.
Watson, J.B. (1913). "Psychology as the Behaviorist Views It." *Psychological Review*, 20, 158–177.
Watson, J.B. (1925). *Behaviorism.* New York: W.W. Norton.
West-Eberhard, M.J. (2003). *Developmental Plasticity and Evolution.* New York: Oxford University Press.
Williams, G.C. (1966). *Adaptation and Natural Selection: A Critique of Some Current Evolutionary Thought.* Princeton, NJ: Princeton University Press.
Wilson, D.S. (1983). "The Group Selection Controversy: History and Current Status." *Annual Review of Ecology and Systematics*, 14, 159–187.
Wilson, D.S. (1997a). "A Theory of Group Selection." *Proceedings of the National Academy of Sciences*, 72, 143–146.
Wilson, D.S. (1997b). "Incorporating Group Selection into the Adaptionist Program: A Case Study Involving Human Decision Making." J. Simpson and D. Kendricks (eds.), *Evolutionary Social Psychology.* 345–386. Mahwah, NJ: Lawrence Erlbaum Publishers.
Wilson, D.S. (1997c). "Introduction: Multi-level Selection Theory Comes of Age." *The American Naturalist*, 150, Supplement, S1–S4.
Wilson, D.S. (1997d). "Altruism and Organism: Disentangling the Themes of Multilevel Selection Theory." *The American Naturalist*, 150, Supplement, S122–S134.
Wilson, D.S. (1998). "Hunting, Sharing, and Multilevel Selection: The Tolerated Theft Model Revisited." *Current Anthropology*, 39, 73–97.
Wilson, D.S. (2002). *Darwin's Cathedral.* Chicago, IL: The University of Chicago Press.
Wilson, D.S. (2008). Personal Communication: 9/18/08.
Wilson, D.S. and Kniffen, K.M. (1999). "Multilevel Selection and the Social Transmission of Behavior." *Human Nature*, 10, 291–310.
Wilson, D.S. and Sober, E. (1994). "Reintroducing Group Selection to the Human Behavioral Sciences." *Behavioral and Brain Sciences*, 17, 585–654.
Wilson, D.S. and Wilson, E.O. (2007). "Rethinking the Theoretical Foundation of Sociobiology." *Quarterly Review of Biology*, 82(4), 327–48.
Wilson, D.S. and Wilson, E.O. (2008). "Evolution 'for the Good of the Group'." *American Scientist*, 96, 380–89.
Wilson, E.O. (1971) *The Insect Societies.* Cambridge, MA: Belknap Press of Harvard University Press.
Wilson, E.O. (1975). *Sociobiology: The New Synthesis.* Cambridge, MA: Harvard University Press.
Wilson, E.O. (1978). *On Human Nature.* Cambridge, MA: Harvard University Press.
Wilson, E.O. (1984). *Biophilia.* Cambridge, MA: Harvard University Press.
Wilson, E.O. (1994). *Naturalist.* Washington, DC: Island Press.
Wilson, E.O. (1998). *Consilience: The Unity of Knowledge.* New York: Alfred A. Knopf.
Wilson, E.O. (2006). *The Creation.* New York: W.W. Norton & Company, Inc.
Wilson, E.O. (2007). Personal Communication: 7/19/07.
Wilson, E.O. (2008). "One Giant Leap: How Insects Achieved Altruism and Colonial Life." *Bioscience*, 58(1), 17–25.
Wilson, E.O. and Hölldobler, B. (2005) "Eusociality: Origin and Consequences." *Proceedings of the National Academy of Sciences, USA*, 102(38), 13367–71.

Index

Printed in the United States of America